Marvelous Microfossils
Creators, Timekeepers, Architects

Patrick De Wever

Foreword by Hubert Reeves

Translated by Alison Duncan

Johns Hopkins University Press
Baltimore

Foreword

This book's magnificent images have awakened one of the fondest memories of my childhood. One day, invited by a geneticist family friend to take a peek through the lens of his microscope, I discovered with awe a teeming world of little organisms. There were so many. They were moving all around. They were colliding like bumper cars at a carnival.

Astounded, I lifted my head. Looking all around me, I asked, "Where are they?"

"There," responded this friend, indicating the gray slide under the microscope tube.

"Where?"

"Right there, before your eyes."

Mystified, I took my time to get used to the idea that all of this extraordinary world was contained in this very small, seemingly unremarkable drop of water. I placed my eye back on the eyepiece, unable to pull myself away from this spectacle.

I felt like I was living in a magical moment. I was discovering a universe! I felt intense gratitude rise in me for this friend who had invited me to see this show. I think my decision come adolescence to choose a scientific profession—and to do research—was largely motivated by this event.

At this time in the early 1940s, the resources available for observing the microscopic world were a far cry from what they are today.

The invention of the electron microscope and all its variations masterfully increased our understanding of the bestiaries of these miniscule places. We continue to discover, with amazement, unbelievably strange forms as we also continue to expand how we identify the paths that life offers inert matter to unfold into existence. This is what this book magnificently illustrates.

For the first three-quarters of its existence, life on Earth existed in a form invisible to the naked eye. Macroscopic life appeared only in the last billion years. The collective mechanisms that transformed inert matter into living matter are still largely unknown to us. Studying these microscopic earthly forms is imperative, as it is particularly promising for revealing these mysterious processes.

This information is equally important for research on extraterrestrial life on the exoplanets that orbit neighboring stars in our solar system.

The author has done a remarkable job, and his work will enrich us for a long time to come. We are indebted to him.

Hubert Reeves
Astrophysicist
Honorary President, Humanité et Biodiversité

Diatoms, Cape
of Good Hope,
Deflandre collection.

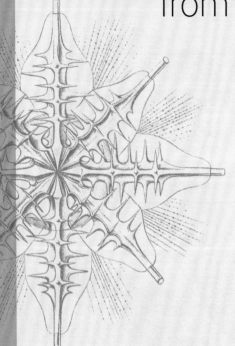

Veritable Stone Flowers
from the Whims of Nature

Throughout history, humankind has collected . . . interesting stones that attract attention by some irregularity in their form or some significant peculiarity in their design or color. Most of the time, the fascination is provoked by an unexpected, unlikely and yet natural resemblance.

. . .

Reflection rightly marvels at the observation that nature, which can neither draw nor paint the likeness of any object, sometimes gives the illusion of having succeeded in doing so, whereas art, whose attempts are always successful, renounces this traditional calling and, inevitably and naturally, is precisely in favor of the creation of blank, spontaneous and unprecedented forms, like those that are abundant in nature.

Roger Caillois,
L'écriture des pierres (*The Writing of Stones*), 1970

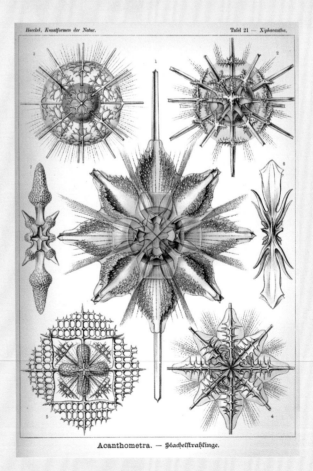

Contents

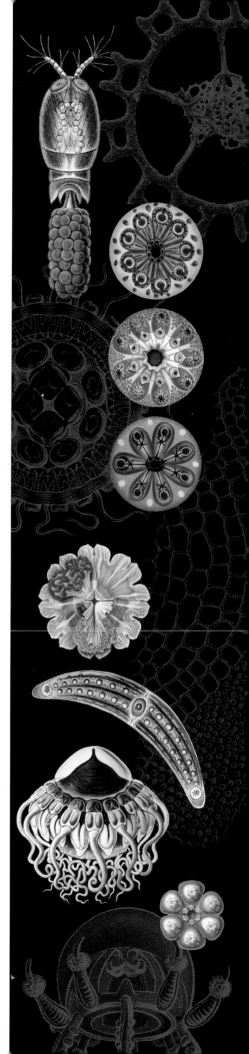

Understanding the Earth

I still count among my most precious memories . . . a pursuit along the flank of a limestone plateau in Languedoc to the line of contact between two geologic layers . . . To an uninformed observer, this quest would have seemed illogical, but it offered my eyes the very image of knowledge, with the difficulties it involves and the joys we can expect.

Claude Lévi-Strauss

Claude Lévi-Strauss invites us to question ourselves.[1]

I quote:

"For is it not because of the myth of the exclusive dignity of human nature that nature itself suffered its first mutilation, to be followed inevitably by other mutilations?

"We began by severing man from nature and setting him up as a sovereign kingdom apart. With this we thought we had done away with the one characteristic that can never be denied, namely that man is first of all a being that is alive. And by closing our eyes to this common feature, the door was opened wide to every outrage and abuse. Never in the course of the past four centuries has Western man been in a better position to realize that by arrogating to himself the right to raise a wall dividing mankind from the beast in nature, and appropriating to himself all the qualities he denied the latter, he was setting in motion an infernal cycle. For this same wall was to be pulled steadily tighter, serving to set some men apart from other men and to justify in the minds of an ever-shrinking minority their claim to being the only civilization of men. Such a civilization, based as it was on the principle and notion of self-conceit, was corrupt from the very start."

---

1. Excerpt from an address given on June 28, 1962, in Geneva during the 250th anniversary celebration of J.-J. Rousseau's birth, "Jean-Jacques Rousseau, fondateur des sciences de l'homme," *Anthropologie structurale deux* (Paris: Plon, 1973), 49–55. ["Jean-Jacques Rousseau, Founder of the Sciences of Man," *Structural Anthropology*, volume 2.]

Introduction

I spent a few decades studying microfossils. My time was divided between field work, both on land and at sea (from 5,600 m in the Himalayas to –5,500 m off the coast of Peru), long hours in the chemistry lab attempting to extract microfossils—attempting, because success was not often found—and long hours with the microscope (binocular and electron). Tedious work, indeed. However, every now and then, very rarely to be precise, a sample revealed well-preserved radiolarians (siliceous planktonic organisms). Then, it was like fireworks of forms, patterns and details. There were thousands of elements, miniscule spines and delicate three-dimensional lace all made of rock crystal. A deluge of wonders. Love at first sight. Of course, I wanted to share this feeling, however . . . These forms are so small that it is not possible for more than one person to look at them at the same time. Yet, to truly enjoy an emotion, you must be able to share it. Who wants to enjoy a fine bottle of wine all alone? From this practical impossibility grew a vague feeling of dissatisfaction, which with time developed into true frustration. It is to compensate for this frustration that I wanted to make this book. For sharing, especially in the case of this beauty, multiplies the joy. I also wanted to show that many other microorganisms are worthy of attention, whether that be for their modest beauty, their usefulness or their role in the environment in which we humans evolved.

I did not attempt an exhaustive presentation of every group of microfossils. I chose the forms that seemed the most strange and beautiful to me. Whether large like stars or immensely small like microorganisms, worlds that are not on our scale make us dream. They also teach us that it is neither the most visible nor the largest that is the most important.

May this book invite you to peer inside stones for the beauty hidden within them.

A Marvelous
Microscopic World

Even if we had to settle for what is immediately within reach and easily seen, we would not lack opportunities to be amazed: the relief of a valley, the delicateness of a flower or the grace of a doe. From our youngest age, we detect a world equally marvelous in its complexity and smallness: the world of insects. Something else we admire, if only because it seems less within our reach and more rare, is snow. It alters our world's usual light, it muffles sounds, it blurs edges, it quickly makes us cold and just as quickly very warm once we are back inside the house. On occasion, snow prompts us to let a few flakes fall on our tongues before we notice that these little white dots seem to have pretty geometric shapes, but they fleetingly vanish under the heat of our gaze. This type of childhood discovery happens again and again throughout our lives as long as we remain a little curious.

This sense of wonder can last a lifetime when we see the microscopic world, both because we are aware of looking at elements that are not easily or directly within our reach and because the forms we see seem unusual and, to put it simply, strange.

From left to right and from top to bottom:

Star-shaped carbonate microorganism measuring only a few micrometers, equal to a few thousandths of a millimeter (nannofossil: *Discoaster* sp.).

Ordinary stellar dendrite snowflake, significantly enlarged by a low-temperature scanning electron microscope. Colorized version to emphasize the central flake.

—

Siliceous plankton (radiolarians: *Tritrabs wortzeli*) from the Mesozoic Era (Jurassic, 145–165 million years ago).

Flake under a scanning electron microscope (SEM). Stellar snow crystal.

—

Star-shaped carbonate microorganism measuring a few micrometers (a few thousandths of a millimeter), formed from marine sediments from the Tertiary Period (a few million years ago). Photo taken with a scanning electron microscope (nannofossil: *Discoaster surculus*).

Snowflakes observed under low-temperature SEM. The image is shown in false color, a practice frequently used to bring out different elements. The original image is black and white.

—

Siliceous plankton from the Mesozoic Era (Triassic, 242 million years ago).

Snowflake seen outdoors with an optical microscope.

What Is
a Microfossil?

Any dead organism, or trace of an organism, that has been preserved in sediment or rock can be called a fossil, no matter how it was preserved or how long ago it died.

Micropaleontology is the study of those fossils that can be seen microscopically. They are millimetric in size. When they are smaller than 50 micrometers (μm, equal to 0.05 mm), they are considered nannofossils, from the Greek *nannos*, meaning "dwarf." Microfossils encompass all fossils with a small size, regardless of whether the living organism from which the remains came was much larger. Therefore, they consist of a microorganism's whole skeleton or of separate skeletal elements or fragments or of fragments (fish teeth and scales, pollen grains, etc.) of larger organisms. Therefore, all groups of organisms potentially fall within the field of micropaleontology. For living organisms, their fossilizable parts are of variable chemical nature: some are silica, others carbonates or phosphates, and others still are formed from non-mineralized organic compounds (chitin, sporopollenin, etc.).

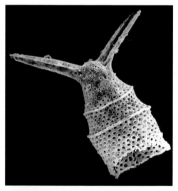

Top, from left to right:
Radiolarian from Turkey. This planktonic organism lived 185 million years ago (Early Jurassic). Size: approx. 0.2 mm.

Part of a nanoscale algae (*Discoaster surculus*). Age: less than 65 million years old. Size: approx. 0.01 mm.

Part of a nanoscale algae (*Calcidiscus leptoporus*). Size: approx. 0.01 mm.

Bottom:
Planktonic foraminifer (*Neogloboquadrina pachyderma*). These organisms prefer cold waters and have been in existence for approximately five million years. The maximum size is close to half a millimeter.

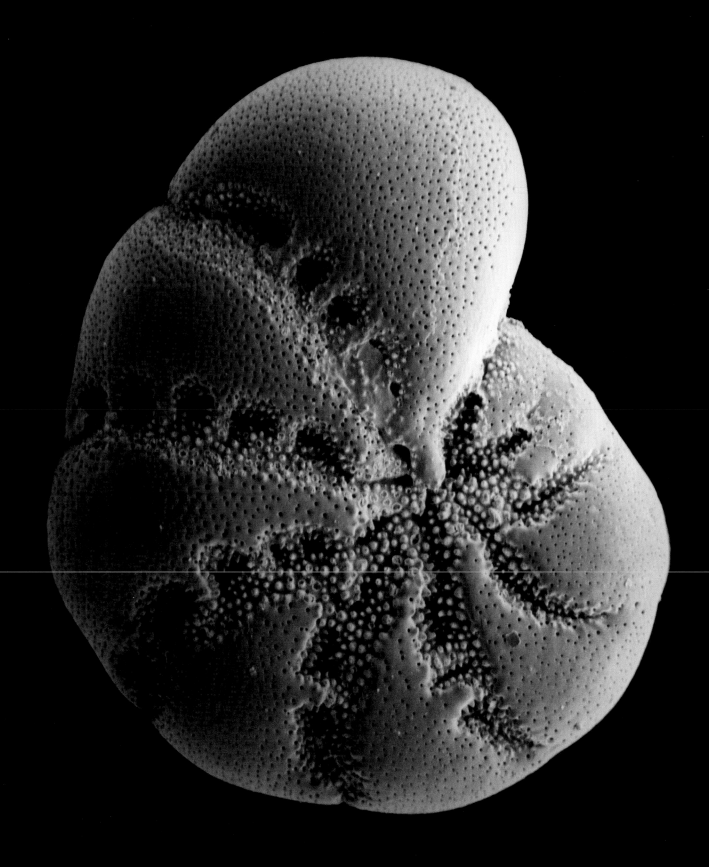

Deep-sea foraminifer:
Elphidium. These
organisms have lived in
the ocean for a few million
years. Approx. size: 0.4 μm.

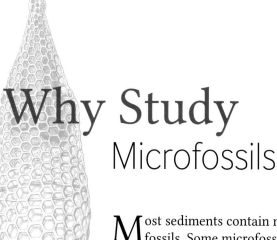

The possession of knowledge does not kill the sense of wonder and mystery. There is always more mystery.

Anaïs Nin

Why Study
Microfossils

Most sediments contain microfossils. Some microfossils are preserved when the sediment has changed into rock after compaction and cementation.

The types of organisms found in sediments depend mainly on the environment, their age and what happened to them while they were buried. In the places where microfossils are the most abundant, as in the sands of backreefs for example, 10 cubic centimeters of sediment can contain thousands of individual microfossils belonging to more than 300 species. Sediment deposits do not occur immediately; they require time. A layer of sediment a few centimeters thick represents hundreds or even thousands of years of accumulation. Even so, on a geologic scale, this level of accuracy is often enough for dating a rock.

Microfossils are therefore an effective, convenient, almost ubiquitous tool for determining the age and environment of the deposits in a layer of sediment. This is why the microscopic world is widely used in academic and industrial fields, particularly for oil exploration and mining research, as well as for land use planning. Indeed, only one small fragment or a shaving from drilling allows a rock to be dated. This avoids the need for core sampling, which is much more expensive and time consuming.

Microfossils are likewise used to look for possible life on other planets. Being so tiny, they could be found even in small samples.

Illustrations of radiolarians (siliceous plankton) by E. Haeckel, published in 1862 (*Die Radiolarien: Eine Monographie* [Radiolaria: A Monograph]).

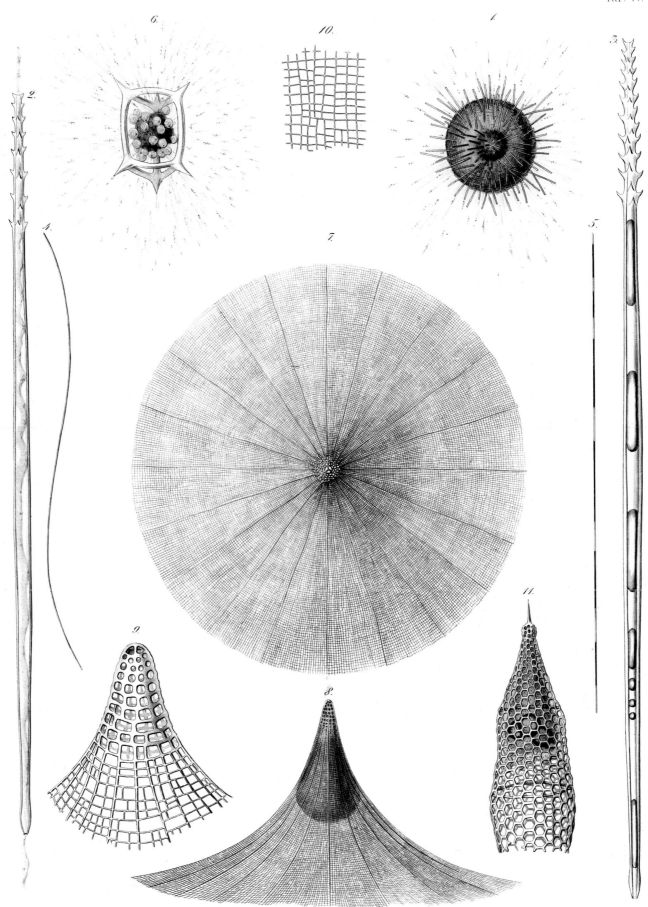

1–5. Aulacantha scolymantha, Hkl. 6. Acanthodesmia Prismatium, Hkl.
7–10. Litharachnium Tentorium, Hkl. 11. Eucyrtidium Lagena, Hkl.

E. Haeckel del.   Wagenschieber sc.

7

10

9

The Study of **Microfossils**

*If science does not take an interest in outrageous things,
it greatly risks missing interesting things.*

Antoine Labeyrie, astronomer

How Are
They Studied?

Following the phenomenon of cementation, microfossils embedded within rocks become an integral part of them. Therefore, in order to observe these microfossils, particular techniques should be used.

Certain fossils can be recognized even in slices. Foraminifers are one such example. In order to study them, the rocks are cut into fine strips of 3/100 of a millimeter. At this thickness, which is the international standard, almost all rocks are transparent. The micropaleontologist's job is then to identify the fossils based on a single slice of rock, no matter the direction in which the sample was cut. It's an art! Given the size of these objects, this work is done under a microscope.

Most microfossils must be observed in all three spatial dimensions to be identified both by their overall form and by their details, such as their spines and surface qualities (perforation type, smooth or rough surface, ornate surface, etc.). Therefore, the challenge is to extract the fossil remains so they can be observed with different types of magnifying devices, including: optical microscopes, stereomicroscopes, scanning electron microscopes (see p. 22), tomography (see p. 24) and so on. Routine work is generally done easily with an optical microscope.

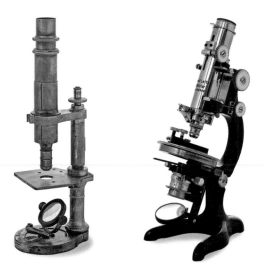

Left: Small upright Nachet microscope, circa 1881.

Right: Leitz microscope, circa 1920. Far more sophisticated, it can analyze objects with controlled lights (polarized, conical beam, etc.).

A thin strip of rock seen under a microscope reveals fossils. Here are nummulites from the Paris Basin from the Lutetian Age (approx. 50 million years ago).

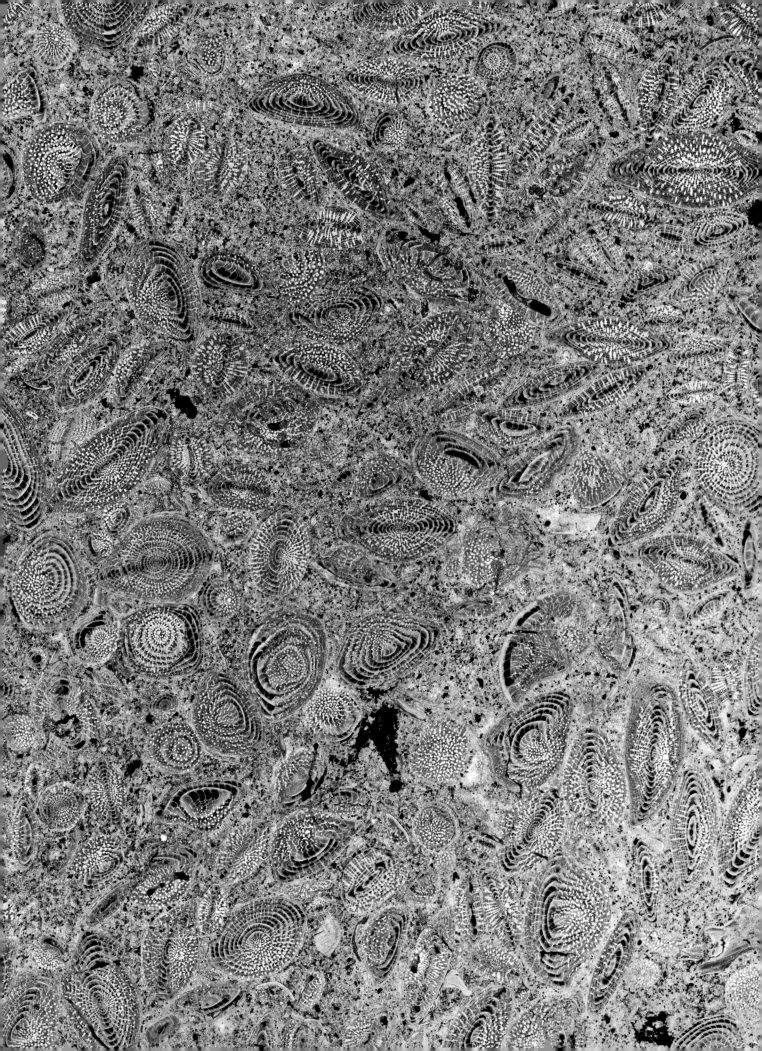

Liberating
Microfossils

*While science reveals the secrets of nature,
what it loses in mystery,
it gains in wonder.*

Paul Carvel

Microfossils that must be extracted from rock pose two problems: their size and their fragility. This usually rules out all mechanical extraction methods. Therefore, extraction is done chemically; we try to dissolve gangue without altering the fossil that we want to study.

Depending on the nature of the matrix and the fossil, the products used can vary, but most of the time acids are used. Acetic acid and hydrochloric acid are used to dissolve limestone. Hydrofluoric acid, a dangerous solution, is used to dissolve the entire rock when it is clayey or siliceous or when fossils are in organic matter or silica (quartz). However, in the latter case for instance, we of course want to dissolve the siliceous gangue without dissolving the fossils, which are also siliceous. To do so, we then try to play with the difference in solubility of quartz crystals, depending on the quality of their crystallization. Chemistry is delicate work!

And the goldsmith's work does not stop there. The undissolved residue must be separated, under the microscope, and the organisms removed without breaking them. Much like the work of a Benedictine clockmaker!

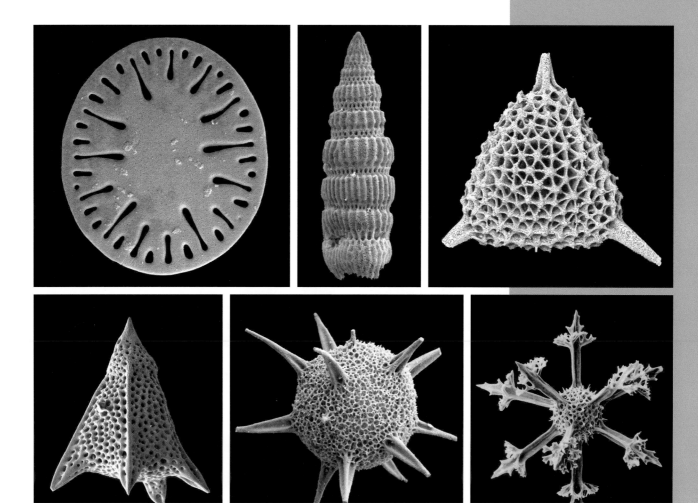

Top, from left to right:
Sclerites from Romania from
the Late Cretaceous Period
(80 million years ago). Photo
taken by scanning electron
microscopy. The organism
measures approximately 0.2 mm.

Radiolarian (nassellaria) from
rocks in Romania. This planktonic organism lived in the
ocean approximately 80 million
years ago. Photo taken by
scanning electron microscopy.
The organism reaches nearly
0.3 mm in length.

Radiolarian (nassellaria,
Alievum superbum) from rocks
in Romania. This planktonic
organism lived in the ocean
approximately 80 million years
ago. Photo taken by scanning
electron microscopy. The organism reaches nearly 0.3 mm.

Bottom, from left to right:
Radiolarian (nassellaria,
Lithochytris vespertilio) from
deep drilling sites in the
southwest Pacific Ocean. This
organism lived in the ocean
approximately 45 million years
ago. Photo taken by scanning
electron microscopy. The
organism is approx. 0.3 mm.

Radiolarian (spumellaria, *Astro-
centrus* sp.) from 242-million-
year-old rocks (Triassic, Anisian).
Photo taken by scanning electron microscopy. The organism
measures approx. 0.3 mm.

Radiolarian (*Cladococcus* sp.)
from deep drilling sites in the
present-day ocean. These
organisms have lived in the
ocean for approx. 12,000 years.

Necessity is the mother of invention.

Plato,
The Republic

Luting with a Turret:
A Useful Technique

Bitumen of Judea, sometimes called lute, is a photosensitive material and a type of natural tar used since ancient times. Ancient people collected it from the surface of the Dead Sea (which the Greeks called Lake Asphaltites) as it continually rose from the bottom of the water. This bitumen was used by the Egyptians to embalm mummies, to caulk ships, to lay paving stones and even to cover the ground of the Hanging Gardens of Babylon. Bitumen was turned into kerosene with an alembic (a still) as Al Razin described in the 9th century.

In the 19th century, people already knew how to extract bitumen from rocks, so it was no longer necessary to import bitumen of Judea as Niépce had done for his photographs.

Bitumen of Judea was once widely used to protect prepared microscope slides from contact with the air. This lute was usually applied with a brush. It makes an impeccable sealant when laid on with a brush affixed to a small lathe (or turret) that deposits the lute in a steady coil as it rotates. The objects to be observed are placed in the center of this circular area, and a strip of glass placed on

this torus protects the fragile object. We then have "luting with a turret," and it is indeed a mounting technique for microscopy and not one of the chapters of the *Kama Sutra*.

This microtableau, which is approximately 1 mm, is a mosaic of colored elements. Each element is a scale from a butterfly wing (certainly not a microfossil).

Radiolaria
Fundort-Platte
mit *100* Formen

A. Elger, Eutin I. H.

Möwe-Expedition
Indischer Ocean
26° 47' Östl. L.
36° 27' S. Br.
4020 m Tiefe

Photo of a prepared slide containing preparations of 100 radiolarians (so small that they are impossible to see in the middle of the prepared slide) from a depth of 4,000 m in the Indian Ocean. The sample is protected by a small coil of bitumen mounted with a turret.

This rosette is a microscopic arrangement of five phytoplanktonic organisms: four dinoflagellates (*Ceratium symmetricum*) encircle a diatom in the center, from the Sea of Japan. Total size: approx. 0.5 mm.

Microtableaus measuring approximately 1 or 2 mm. Each rosette is made of hundreds of elements, each of which is a skeleton from a small siliceous alga: a diatom.

Thinking, analyzing, inventing are not anomalous acts;
they are the normal respiration of the intelligence.

Jorge Luis Borges

The Scanning
Electron Microscope

The term "microscope" generally refers to an optical device. This has been the case for a long time. In the middle of the 20th century, instruments were developed to analyze the behavior of electrons. An object would be bombarded with electrons, and the instrument would scan this object line by line. Accordingly, this instrument was named the scanning electron microscope.

Depending on the geometry of the observed object, projected electrons are deflected to varying degrees. By making the volume of the object stand out, it is thus possible to obtain images in relief.

The resulting images, corresponding to stronger or weaker electrical charges, are only in black and white. For the sake of aesthetic or pedagogical concerns, these images are sometimes artificially colored.

This tool has considerably increased our observation capabilities; it has increased them by a factor of a thousand!

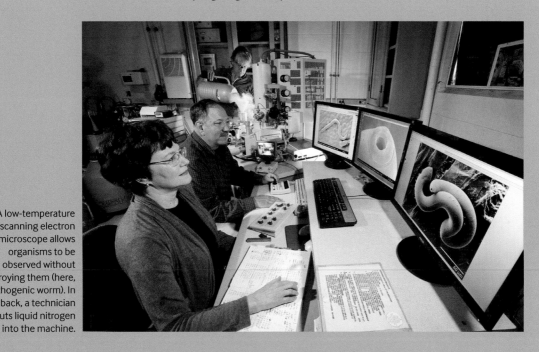

A low-temperature scanning electron microscope allows organisms to be observed without destroying them (here, a pathogenic worm). In the back, a technician puts liquid nitrogen into the machine.

Electron microscope
from the 2000s.

The first scanning electron
microscope in France.
Geology laboratory, the
French National Museum
of Natural History, photo
taken by Robert Doisneau.

*For nothing is covered up that will not be uncovered,
and nothing secret that will not become known.*

The Bible, NRSV, Matthew 10:26

Observing
in Three Dimensions

In the natural sciences, a study generally begins with a morphological description of an object's exterior and interior. The internal structures are important for understanding the biology of and relationship between living things. For these reasons, since the beginning of the 20th century, paleontologists have been familiar with tomography techniques (which means sectional imaging, from the Greek *tomos*, meaning "section"). For a long time, these techniques involved gradually eroding the specimen using an abrasive and drawing, step by step, what became visible. The drawings were then superimposed on transparent sheets, making it possible to reconstruct the image in three dimensions. An alternative to sanding fossils was the serial ultrathin sectioning technique, which was generally done after embedding the specimen in resin. In either case, once the experiment was over, nothing remained of the fossil itself. The analysis was destructive. Therefore, these techniques were only used on relatively common specimens, never on a rare or unique specimen.

The emergence of computed tomography was a major breakthrough in the 1980s. This type of tomography can reproduce the volume of an object from a series of measurements made in successive installments without destroying the object. The result is a reconstruction of its external and internal anatomy. From that we can reconstruct the object through three-dimensional printing.

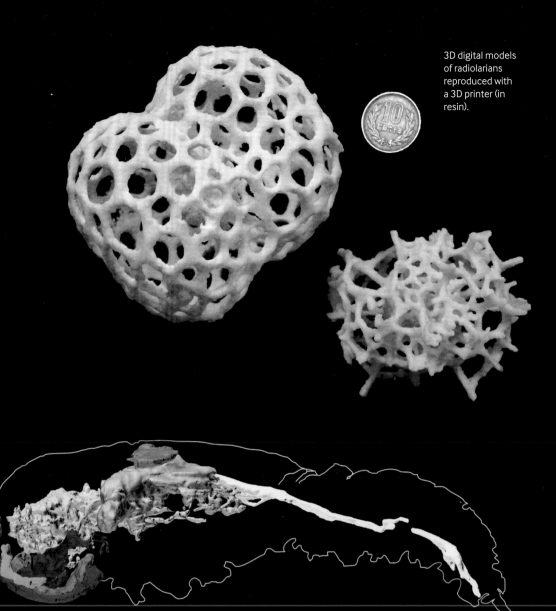

3D digital models of radiolarians reproduced with a 3D printer (in resin).

Lateral view of *Voulteryon parvulus*. Reconstruction of the internal parts of a tiny lobster from 165 million years ago from La Voulte-sur-Rhône, France. Because this fossil is contained in a carbonate nodule, only a small part of its carapace can be seen. A CT scan allows the internal parts, which have been colored, to be reconstructed without destroying the fossil. The organism is approximately 10 mm in length.

Microtomography: image of flint taken using a synchrotron revealing numerous embedded spicules from sponge fossils (rods and small spears). Flint from the Cretaceous Period (Cenomanian, 100 million years ago, Charente-Maritime, France). Data acquired at beamline ID19, European Synchrotron Radiation Facility, Grenoble, France. Method: propagation-based phase-contrast microtomography. Volumetric pixel size: 11.8 μm.

Observing the Microscopic
by Satellite

Large animals often impress our eyes as children, yet it is micro-organisms that can be seen from several tens of kilometers away.

Sometimes plankton experience an explosive increase in population, called a bloom. These blooms are temporary. Sometimes they are visible from an altitude of up to 10 kilometers. They stand out by discoloring the water, creating a "stain" that can spread up to 7,200 square kilometers and that can contain between 10 and 115 million cells per liter. Blooms are caused by different groups of organisms that make the water take on a milky, turquoise, greenish blue or even red color. In different legends and mythologies, some red tides, which are algal blooms caused by dinoflagellates, have been described as blood. Such phenomena have been documented in various places, such as off the coast of Brittany in France, in the fjords on the Norwegian coastline and in the Black Sea. These blooms can be fossilized as thin layers of organisms that include only one species. We can thus have a record of the seasons, a true snapshot of geologic time.

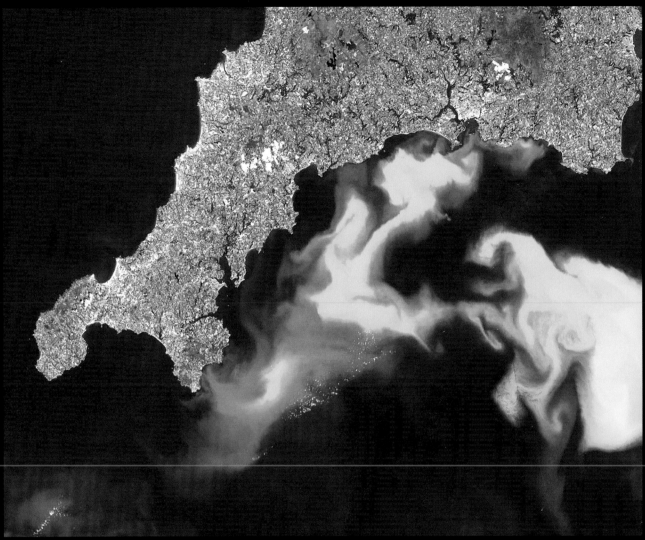

These explosions of life (blooms) produce billions of nannoplanktonic organisms in the water column. Despite the nanoscopic scale of each individual organism, this bloom was spotted from afar by the light it reflects. It was caused by the algae *Emiliania huxleyi*, a coccolith. Image taken from the Landsat satellite on July 24, 1999.

I was told that in the country of Valois, near a place named Venteul, there was a large quantity of petrified shells, . . . similar to those of the Ocean Sea [the tropical Atlantic].

Bernard Palissy

Paris
in the Tropics

Only geologic time makes it possible to imagine that the present-day mountains were formed from the ocean floor and, because of variations in sea level, that most of today's inhabited regions were formerly under water. Thus, Paris was built from myriad minuscule skeletons of organisms that lived on the bottom of a warm, shallow sea.

Indeed, 45 million years ago, a clear sea covered immense surfaces, including most of the countries surrounding the Mediterranean today. The climate was tropical back then since the latitude of the Paris Basin was around 40° north. The water and its depths were teeming with small organisms called nummulites (meaning "stone coins"). These small cylinder-shaped gastropods were unicellular organisms in the shape of a small coin, hence the name.

When they died, these organisms settled on the ocean floor. The sand that formed consisted of only these complete or fragmented remains. Over time, and with the pressure of overlying sediment, the sand hardened and became rock. The stone that formed is pale beige, and it contains small grains that match the color of these organisms' skeletons. The material is solid but can be carved quite easily, and since it is very homogeneous, it is used as building stone. In fact, Paris was built with this high-quality material, and it is the stone's unique color that gives a pleasant cohesiveness to the city's buildings.

Notre-Dame de Paris Cathedral seen from the southeast. The limestone used for the building was formed from sediment composed of fragments of microscopic organisms that lived in the area nearly 45 million years ago (Lutetian).

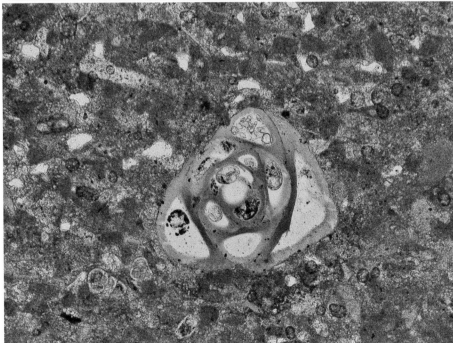

Image of a thin slice (0.03 mm thick) whose transparency shows the wealth of very tiny fossils and a few larger ones, such as this miliolid. The column statues of Notre-Dame de Paris, Notre-Dame d'Étampes and Chartres Cathedral are carved out of this type of limestone. The largest microfossil measures approx. 0.5 mm.

Within the capacity of wonder, there lies one of the secrets of life.

Sylvain Tesson,
Éloge de l'énergie vagabonde
(In praise of roaming energy)

Fossils or Minerals?
Organic Remains or Unrealized Organisms?

The nature of fossils has long been debated. In 1819, Karl von Raumer, a professor of mineralogy at the University of Erlangen, declared that plant fossils made of Silesian coal "are plant embryos nestled within the Earth, never reaching the stage of birth." Therefore, for him, plant fossils are not ancient organic remains but unrealized organisms!

The idea has long persisted that rocks and/or ores were formed in the past and continue to form, much like living things. This mythology associating life with ore was tied to the idea of Mother Earth: mining galleries were likened to the life-giving uterus of Mother Earth. As the Romanian philosopher Mircea Eliade wrote in 1977, miners were considered to be involved "in the development of underground embryology: they precipitate the growth of ores, they collaborate in nature's handiwork, helping nature to give birth sooner." Mineral substances contributed to the idea of the sacredness of Mother Earth.

The concept of vital force was evoked in the 14th century by Jean de Mandeville regarding diamonds: "And they grow together, male and female. And they be nourished with the dew of heaven. And they engender commonly and bring forth small children, that multiply and grow all the year. I have often-times assayed, that if a man keep them with a little of the rock and wet them with May-dew oft-sithes, they shall grow every year, and the small will wax great" (*Voyage autour de la Terre* [*The Travels of Sir John Mandeville*]).

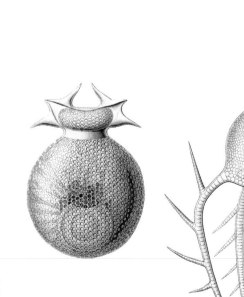

Illustrations of radiolarians (siliceous plankton, Phaeodarea) by E. Haeckel from his published work *Kunstformen der Natur* (*Art Forms in Nature*), Plate 1, 1899–1904.

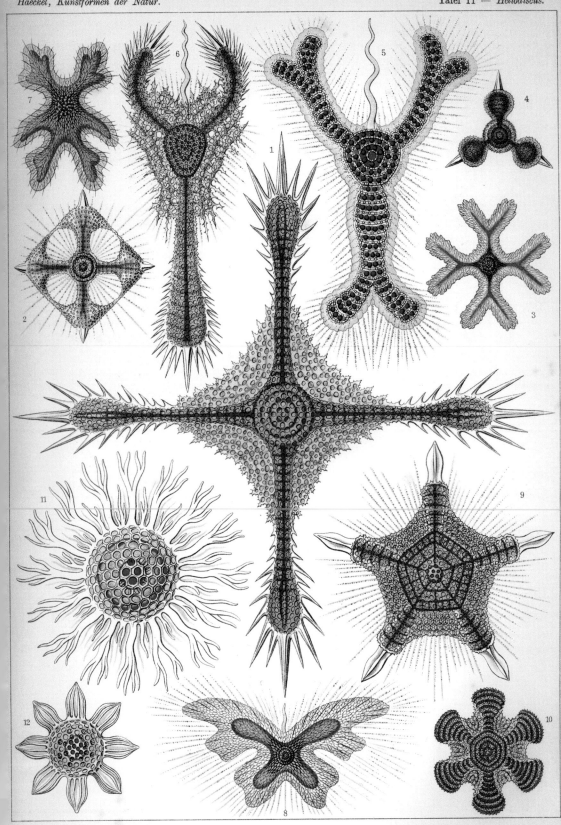

Discoidea. — Scheiben-Strahlinge.

A flash of lightning . . . then night!—Fleeting beauty
Whose glance suddenly brought me back to life,
Will I only see you again in eternity?

Baudelaire,
"à une passante"
("To a Passerby")

Establishing
the Origin of Fossils

Philosophy and religion have long influenced the understanding of what fossils are. Likewise, in spite of explanations proposed since the Renaissance, what fossils are has been a source of passionate disagreement for centuries.

Even if most scholars were in agreement in the 18th century, for some the nature of fossils remained open for debate: are they remains of organisms? Whims of nature? Shells abandoned during pilgrimages? For example, in 1752, Élie Bertrand maintained that fossils, "figurative stones," had mineral origins.

Jean-Étienne Guettard, having published in "different fields of the sciences," demonstrated a broad range of interests, from meteorology to physiology and so on. He "answered Bertrand's daydreams" by explaining the various fossilization and petrification processes of organic bodies. He thus put an end to the age-old controversy over the origin of fossils in a 1759 article titled "On the Accidents of Shell Fossils Compared to Those That Happen to Shells Now Found in the Sea." In passing, he introduced the major principle of actualism. This principle was taken up by Buffon, but it would be defined later and attributed to Lyell.

In paleontology, Guettard emphasized the importance of comparative anatomy for the identification of fossils: "Comparative anatomy is not yet advanced, especially with regard to skeletons, to ensure all the time and all the clarity this area of study demands and that it is likely to receive" (1768). He is thus a precursor of Cuvier and Geoffroy Saint-Hilaire, who revolutionized the study of evolution.

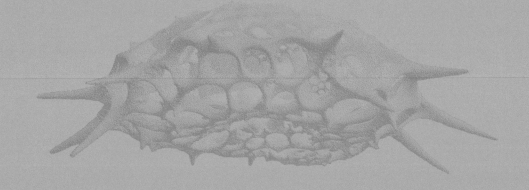

Siliceous plankton, radiolarian (*Parasaturnalis*) from Turkey from the Jurassic Period (180 million years ago). Size at the largest dimension: approx. half a millimeter.

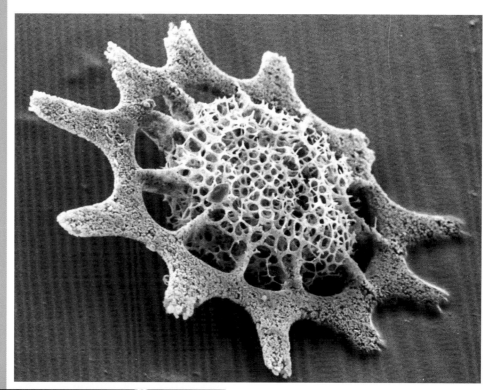

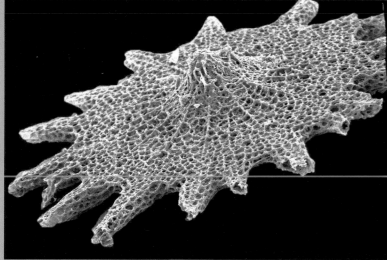

Siliceous plankton, radiolarian (*Citriduma*) from Turkey from the Jurassic Period (180 million years ago). Size at the largest dimension: nearly 1 millimeter. Detail of one spine and the corresponding aperture.

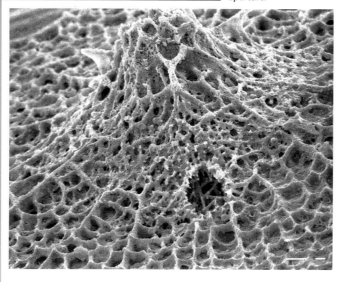

Discovery often means simply the uncovering of something which has always been there but was hidden from the eye by the blinkers of habit.

Arthur Koestler,
The Act of Creation

Discovering the Earth
under Water

The first navigators and explorers of distant lands were mostly interested in geographical surveys of the coasts. However, some naturalists of the 17th and 18th centuries filled Europe's cabinets of curiosities with organisms visible to the naked eye. At the beginning of the 19th century, knowledge of the marine world grew with the growing numbers of exploratory voyages to the southern lands (*Terra Australis*). Nicolas Baudin commanded the French expedition in 1801. A much better-known expedition, the voyage of the *Beagle*, was launched in 1831. It was an English expedition to circumnavigate the world. Aboard was the young Charles Darwin, who had training as a naturalist in zoology, geology, botany and so forth.

The emerging scientific field of oceanography converged with geography through pioneers such as the German explorer Alexander von Humboldt. The microscopic world and ocean floors remained virtually unknown at this time as they were technically inaccessible to scientists. Because of this, they were considered unsuitable for life. And yet . . .

Top center photo: Siliceous plankton, radiolarians, from rocks in Turkey dating from the Triassic Period (Carnian, approx. 230 million years ago). Photo taken by scanning electron microscopy. Size: approx. 0.4 mm.

Top and middle right photos: A few forms of siliceous plankton, radiolarians, from a rock sample in Turkey dating from the Jurassic Period (Pliensbachian, approx. 185 million years ago). Photos taken by scanning electron microscopy. Size: approx. 0.3 mm.

Four middle and bottom left photos: A few forms of siliceous plankton, radiolarians from rocks in Sardinia dating from the Cretaceous Period (Valanginian, approx. 135 million years ago). Photos taken by scanning electron microscopy. Size: approx. 0.3 mm.

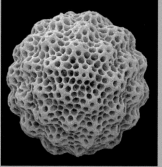

Technical invention is a product of man alone and not of his basic needs, but of his dreams, that is, of his true desires.

Denis De Rougemont

Oceanographic Expeditions
of the 19th Century (1872–1876)

Oceanography was recognized as a science in 1872 when two Scots, Charles W. Thompson and John Murray, set in motion the *Challenger* expedition. Carried out from 1872 to 1876, the campaign began to reveal the incredible wealth of the oceans. During its voyage, the boat traveled more than 120,000 kilometers (approx. 74,565 mi) across the Atlantic, Southern, Indian and Pacific Oceans. The main goal was to study pelagic animals and to understand how water circulated in the oceans. The expedition gave scientists a sense of the geography of seabeds with great ocean basins, discovered the Mariana Trench and proved the existence of the Mid-Atlantic Ridge. It also discovered fields of metallic nodules.

The result was a report that, among numerous discoveries, cataloged 4,000 unknown animal species and created a map of ocean sediments. Much of the new knowledge concerned the microscopic world. This left just as much of a mark on the scientific world as on the artistic world.

Painting of the H.M.S. (His/Her Majesty's Ship) *Challenger*, 1858, by William Frederick Mitchell.

Illustrations of radiolarians (siliceous plankton) published by E. Haeckel in his report on the mission carried out by the ship H.M.S. *Challenger* (*Report on the Scientific Results of the Voyage of H.M.S.* Challenger *during the Years 1873–76*, volume XVIII, 1887).

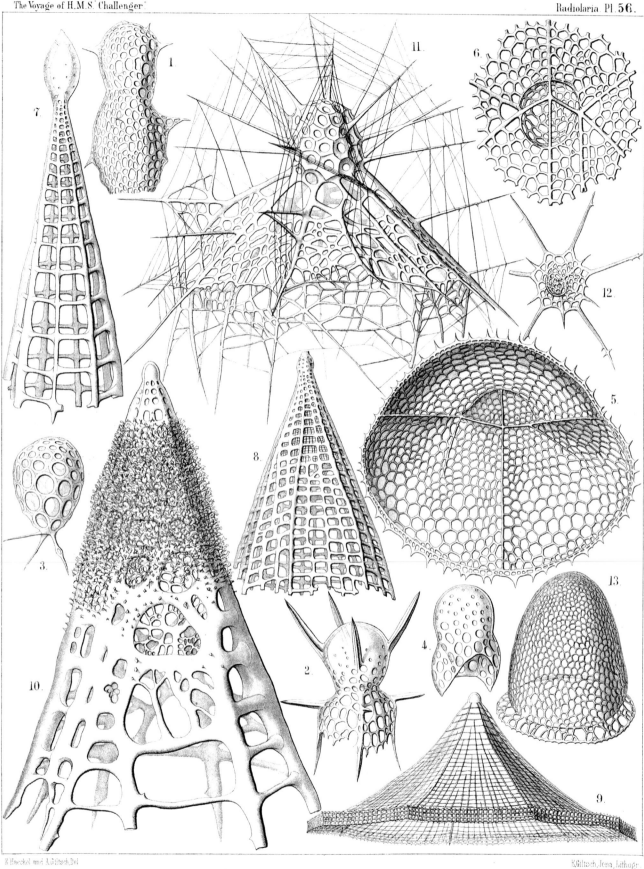

E.Haeckel und A.Giltsch,Del. E.Giltsch,Jena_Lithogr.

1.2. SETHOMELISSA , 3.4.PSILOMELISSA , 5. PENTAPHORMIS , 6. HEXAPHORMIS,
7. CEPHALOPYRAMIS, 8.9. SETHOPYRAMIS, 10. PLECTOPYRAMIS,
11.12. ARACHNOCORYS, 13. SETHOCEPHALUS.

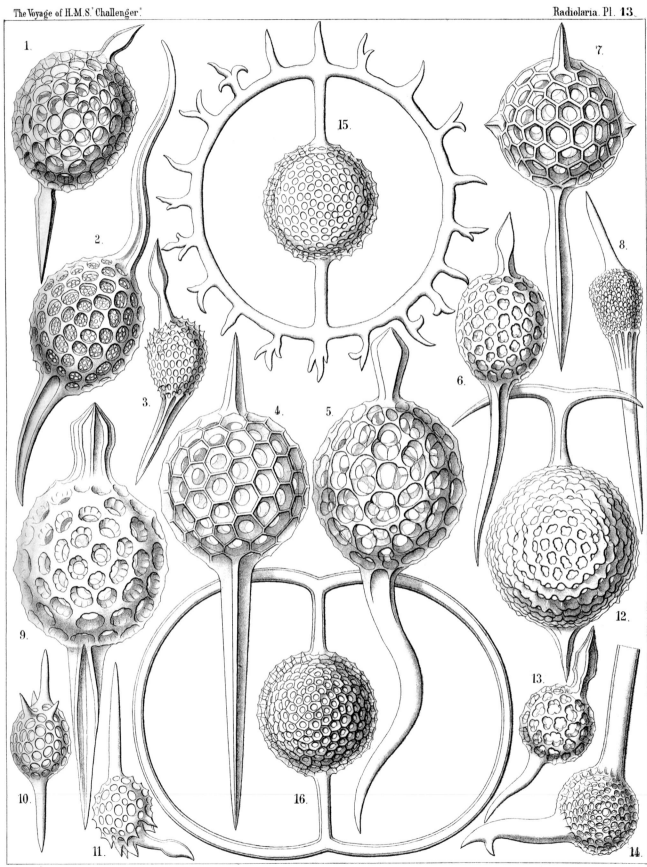

E.Haeckel and A.Giltsch,Del.

E.Giltsch,Jena,Lithogr.

1–14. XIPHOSTYLUS , 15.16. SATURNALIS.

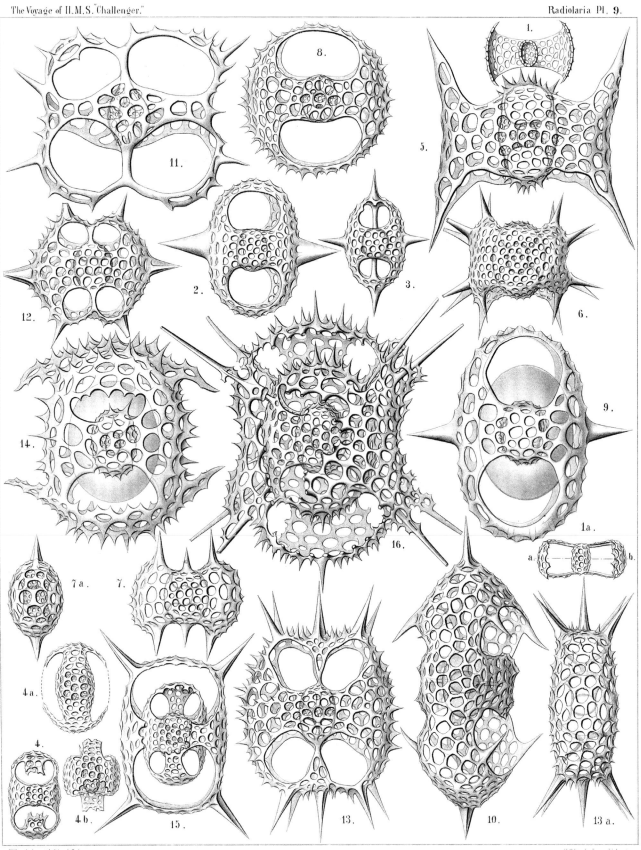

1-4. TRIZONIUM, 5-7. AMPHIPYLE, 8-10. TETRAPYLE,
11-13. OCTOPYLE, 14-16 PYLONIUM.

Terra

Incognita

In the middle of the 20th century, when humans had already set foot on the moon, most of the ocean floor still remained *terra incognita*.

In 1968, the Deep Sea Drilling Project was launched with a specialized vessel, the *Glomar Challenger*. The objective was to collect samples from mid-ocean ridges and abyssal plains to determine their nature and age. Over 15 years, the *Glomar Challenger*, with its 43-meter-high derrick (approx. 141 ft), drilled 1,092 boreholes before being replaced in 1984 by a more efficient vessel, the *JOIDES Resolution*. This boat had the latest deep drilling technologies. It could maintain a stationary position

thanks to 12 computer-controlled engines using satellite positioning information. It also had unequaled investigation capabilities. At 7,000 meters (approx. 4.35 mi) under the water, it was able to drill up to 2,000 meters (approx. 656 ft) into sediment and rock on the ocean floor, a true technical feat.

These programs made it possible to create a map of the ocean floor and collect a large amount of data useful in geology, biology, physics and chemistry. Knowledge within the field of micropaleontology made a huge leap forward, even more so because manufacturers had just discovered its usefulness.

The drilling vessel *JOIDES Resolution* with its derrick that allows it to drill the bottom of the ocean, several thousands of meters below the surface of the water.

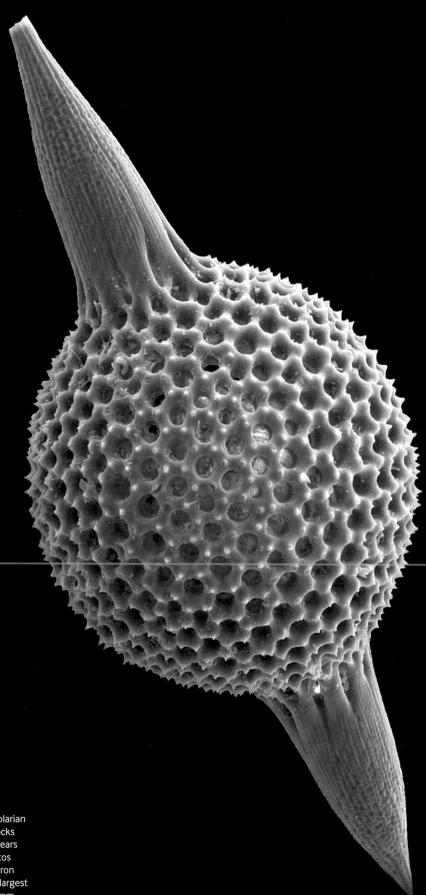

Siliceous plankton, radiolarian (*Spongatractus*) from rocks dating back 45 million years (Lutetian, Tertiary). Photos taken by scanning electron microscopy. Size at the largest dimension: approx. 0.4 mm.

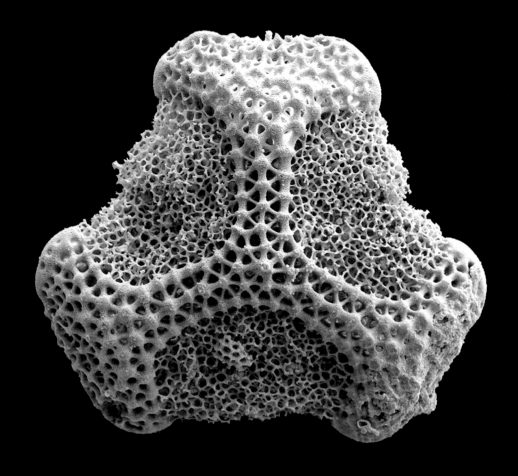

Siliceous plankton, radiolarian (*Halesium*) from rocks in Romania dating from the Cretaceous Period (Coniacian), approx. 88 million years ago. Photos taken by scanning electron microscopy. Size at the largest dimension: approx. 0.4 mm. (Left, detail of the central portion showing the regularity of the arrangement.)

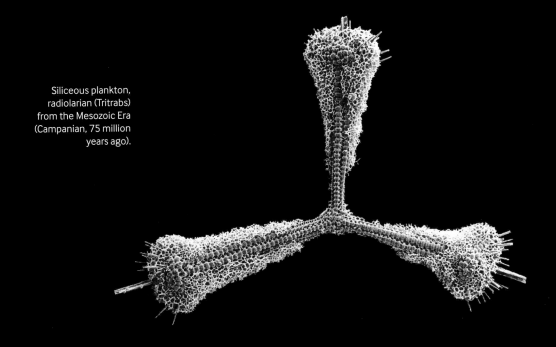

Siliceous plankton, radiolarian (Tritrabs) from the Mesozoic Era (Campanian, 75 million years ago).

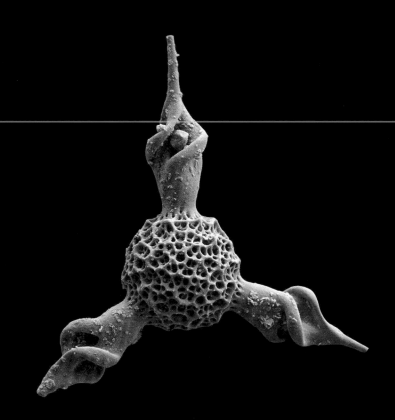

Siliceous plankton, radiolarian from rocks in Turkey dating from the Triassic Period (Carnian), approx. 230 million years ago. Photos taken by scanning electron microscopy. Size at the largest dimension: approx. 0.5 mm.

New Oceanographic
Discoveries

A century after the major oceanographic expeditions of the last part of the 19th century, the international scientific community mobilized around an ambitious program of deep sea drilling with its ships *Glomar Challenger* and *JOIDES Resolution*.

As with the oceanographic expeditions of the previous century, a whole microscopic world of fossils was discovered. Certain organisms previously considered morphologically stable over time proved to be valuable tools for dating sediments. From this point forward, we could determine the age of the ocean floor, thanks to the microorganisms found in the sediments in contact with the basalts emitted during seafloor spreading. We thereby discovered that the farther we went in either direction from the Mid-Atlantic Ridge, the older the sediments. This observation of symmetry proved the theory of spreading. The two edges of the ridge have been moving apart for more than 100 million years, therefore demonstrating continental drift!

In addition, the oldest ocean floor is only 180 million years old, nothing next to the planet's age of 4.5 billion years. This attests to the fact that the oceanic crust regenerates and therefore also moves.

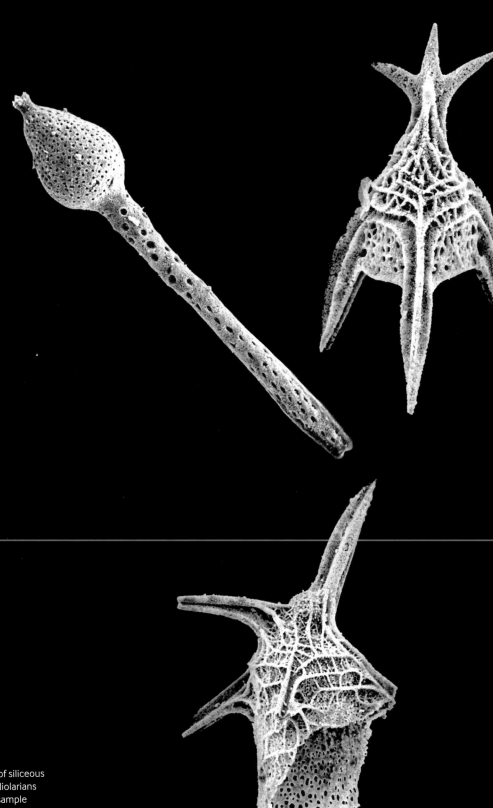

A few forms of siliceous plankton, radiolarians from a rock sample from Turkey dating from the Jurassic Period (Pliensbachian), approx. 185 million years ago. Photos taken by scanning electron microscopy. Sizes range from 0.2 to 0.5 mm.

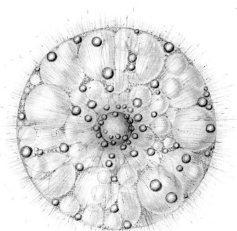

Protists
Unite

Unlike bacteria, which are unicellular nucleus-free beings, protists have a nucleus and organelles. They appeared more than a billion years ago as a result of combinations of organisms composed of genetically distinct cells (called chimeras), through the integration of different bacteria and archaea.

Because they are small, most protists can be seen only under the microscope, but sometimes certain types are visible to the naked eye. This is the case with certain foraminifers, called nummulites, which form the 50-million-year-old nummulite limestone (known as "pierre à liards" [stone coin] after the small French coin no longer in use) in the Laon region of France.

Some protists are capable of photosynthesis (autotrophs) and feed on light, while others feast on other organisms (heterotrophs). However, many are symbiotic and can, in fact, adapt to various modes of nutrition, easily tolerating changes in the environment.

Over time, some protists aggregate to form pluricellular beings. Indeed, their polyvalent cells have specialized for certain functions (muscular, digestive, nervous, reproductive, etc.), forming beings that can no longer live as a single multifunctional cell as they had before. These are called metazoans, among which are included plants and animals.

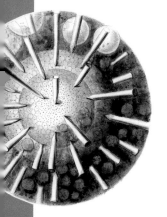

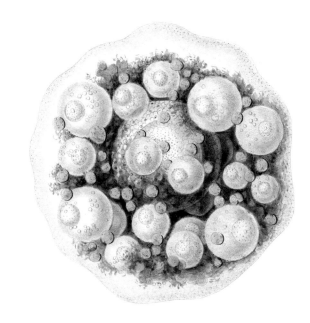

Illustrations from
Die Radiolarien: Eine Monographie (Radiolaria: A Monograph), Berlin: Altas, 1862.

Unidentified organism from the South Pacific Ocean. Image taken by phase-contrast microscopy. We find a central part (detail at the top left) that seems to be a radiolarian with four long extensions, at least three of which carry a cytoplasmic sleeve in which several types of elements are visible (as shown in the detail at the top right); at least some are symbiotic photosynthetic algae. Size measured from the upper extremity to the lower extremity: approx. 2 mm.

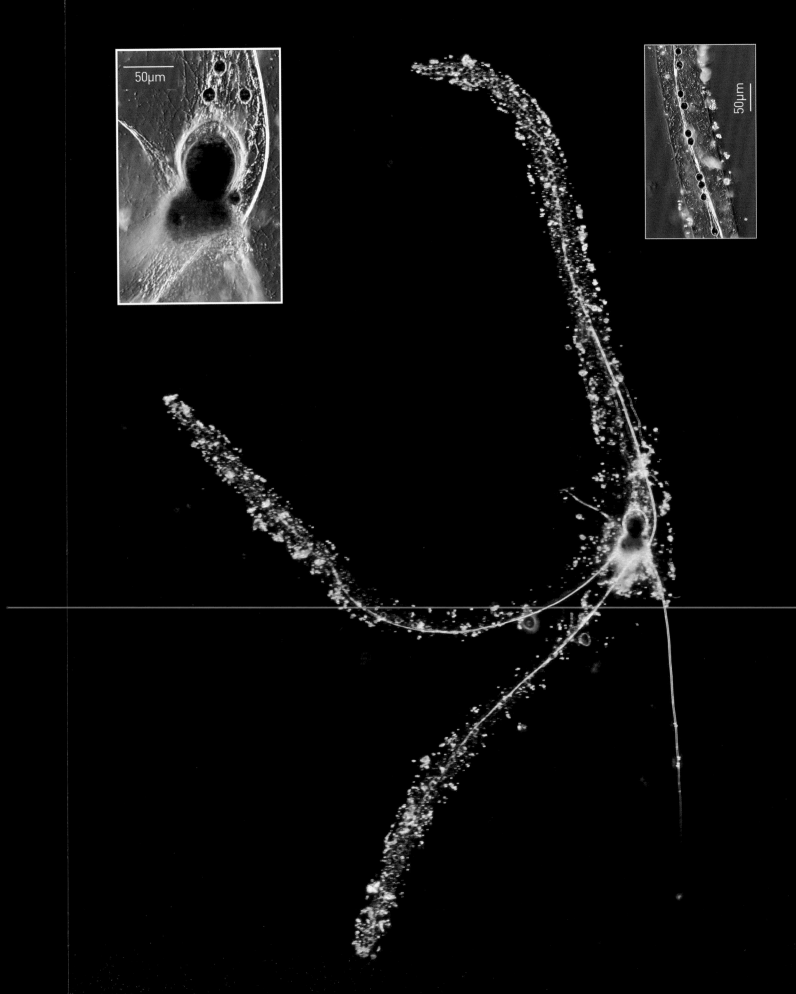

50μm

50μm

Microfossils throughout the Geologic Ages

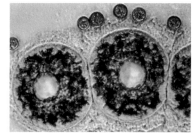

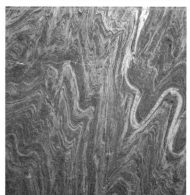

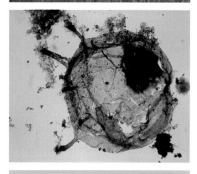

Worlds are born and die,
that which is sea can become land.
Everything changes over time.
Wisdom begins in wonder.

Aristotle

Earth, Our Nourishing Mother,
at 46 Years Old

Mother Earth is like a 46-year-old person for whom each year lasts 100 million years. Like us, she has a very bad memory of her early childhood, which is still very poorly understood.

Her noteworthy events are lost, crushed and recycled by geologic processes. We got a first glimpse of her when she was eight, but this snapshot is discolored, faded, and only blurry outlines remain visible. The earliest records of primitive life in her oceans were visible when she was 11 years old, but they developed so slowly that life-forms with hard parts, such as trilobites, did not appear until she passed canonical age, which is to say, the age of 40.

Her continents remained devoid of any visible life until she turned 42 years old. But her forties marked the beginning of flourishing activity, as her last four years were very busy.

Indeed, when she was 43 years old, sharks appeared in her oceans, insects in her air and forests and reptiles on her land. At age 44, dinosaurs pranced around on her lands and trilobites were already extinct from her waters.

The dinosaurs died eight months ago, their place taken over by mammals. For only two weeks, great anthropoid apes made their appearance.

Last week the glaciers of the Ice Age advanced; they withdrew about 50 minutes ago. But four hours have passed since a new species—which has christened itself *Homo sapiens*—began to hunt other animals, and during this last hour, it invented agriculture and settled in one place. A quarter of an hour ago, Moses led his people safely through a crack in the Earth's surface, the Red Sea. Around five minutes later, Jesus was preaching on a mountain a little farther away, along a large fault line called the Dead Sea Transform, in which settled the River Jordan as it left the Red Sea, ran along the Sinai Peninsula, and, as it continued, created small basins called the Dead Sea and the Sea of Galilee.

One minute before Mother Earth turned 46, humans began their industrial revolution. During that minute humankind tremendously multiplied its skills and overexploited the planet for its ores and sources of energy.

The Venus of Willendorf depicts the Mother Goddess, or Mother Earth, bearing all the attributes of fertility. Found in 1906 in Austria, dating from the Paleolithic.

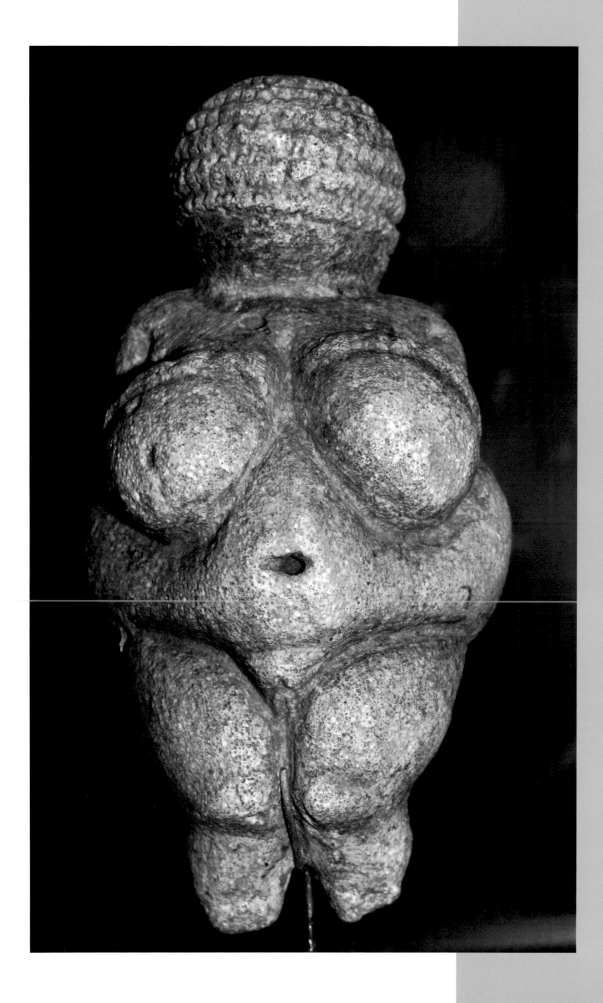

A Microscopic World
of Primary Importance

A thin layer of white or light-colored ooze covers one-sixth of the Earth's surface, mostly on the ocean floor. This film seems insignificant; however, seen under the microscope, this sediment becomes impressive. It reveals countless minuscule skeletons that resemble trumpets, shuttlecocks, water wheels, flasks, balls, sieves, spaceships or even Chinese lanterns. Some of these figures show a glassy luster, others a saccharoid white, while others still a strawberry or orange color, candy pink or apple green; the possibilities go on. This attractive world of microorganisms has existed since the dawn of time, hundreds of millions, even billions, of years ago. Moreover, this world is of primary importance both at the biological level, for its diversity and interactions with other organisms, and at the geologic level, as it is the origin of, for instance, all limestone.

Illustrations of radiolarians (siliceous plankton, *Acanthophracta* sp.) by E. Haeckel from his published work *Kunstformen der Natur* (*Art Forms in Nature*), 1899–1904.

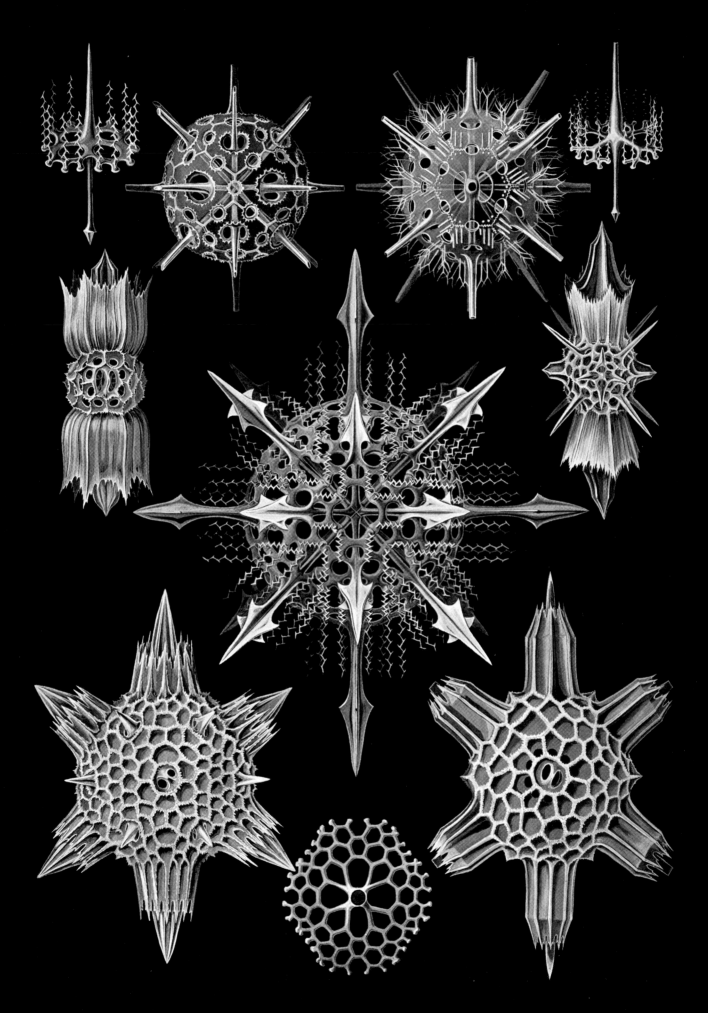

Every evening, hoping for an epic future,
The phosphorescent azure of the tropic sea
Enchanted their sleep with a golden mirage;

Or leaning over the bow of their white caravels,
They watched ascending in the unknown sky
From the depths of the ocean new stars.

José Maria de Heredia,
"Les conquérants"
("The Conquerors")

Haeckel:
The Art of the Introduction

Most of Ernst Haeckel's (1834–1919) scientific work was focused on "invertebrates" (sponges and jellyfish) and the microscopic world (notoriously, radiolarians). The works that made him famous were mainly his marine biology monographs, including ones on radiolarians (1862, 1887). He named more than 3,500 species collected during the *Challenger* expedition (see "Oceanographic Expeditions," p. 36).

Haeckel proposed the idea of the common origin of all organisms and introduced tree diagrams to represent evolution in biology. He also put forward the idea that the origin of life was influenced by physical and chemical factors, such as light and the presence of oxygen, water and methane. Haeckel's ideas had a significant influence on the history of evolutionary theory. He is considered the father of ecology and, incidentally, the creator of this term in 1866. For him, "oecology" (according to his own spelling) referred to the study of the interrelationships of living organisms.

Some of this German naturalist's works achieved enormous print runs, which won him the favor of the general public but not that of all of his colleagues. They criticized him for extrapolating from hypotheses that were insufficiently supported by facts. In the last years of the 19th century, Ernst Haeckel turned more and more toward philosophy. He asserted the fundamental unity of the organic and inorganic and of mind and matter (monism).

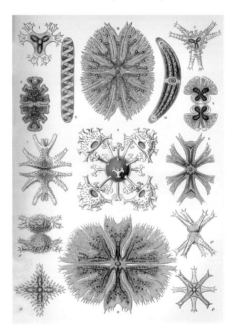

Illustrations of radiolarians (siliceous plankton, *Spyroidea* sp.) by E. Haeckel from his published work *Kunstformen der Natur* (*Art Forms in Nature*), 1899–1904.

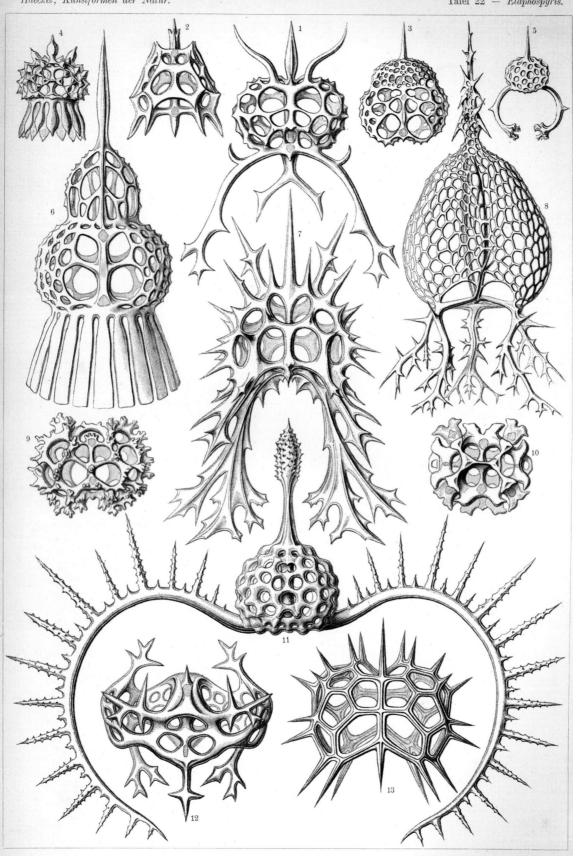

Spyroidea. — Nüßchenstrahlinge.

*So much energy is spent
so that everything will be very still.*

Bernard Werber,
La révolution des fourmis
(The revolution of the ants)

The Economies
of Energy

A large expenditure of energy is required to remove chemical elements that are generally found at levels of subsaturation from water (see "Expending Energy to Make Skeletons," p. 90). This energy is drawn from the surrounding environment, so it is preferable for the environment to be rich in food. If it is not, then organisms that have thin skeletal elements have an advantage over those with massive, thick skeletons.

Over time, this dietary constraint has selected the most efficient organisms, which explains why at various times during the geologic ages dense skeletons became lighter. Skeletal walls became less thick, more perforated, and the cylindrical spines became triradial. Clearly, lighter skeletons were "more successful." The organisms with a better volume-to-surface ratio had an advantage. They had more cytoplasmic material and less of a need for skeletal material. Thus, between the Eocene (around 40 million years ago) and the Quaternary (present day), the average weight of radiolarian skeletons decreased by 75%.

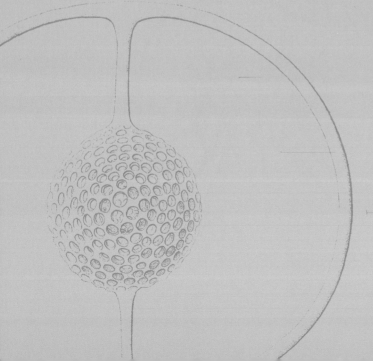

E. Haeckel's illustrations of radiolarians (siliceous plankton) from his report on the mission carried out by the ship H.M.S. *Challenger* (*Report on the Scientific Results of the Voyage of H.M.S.* Challenger *during the Years 1873–76,* volume XVIII, 1887).

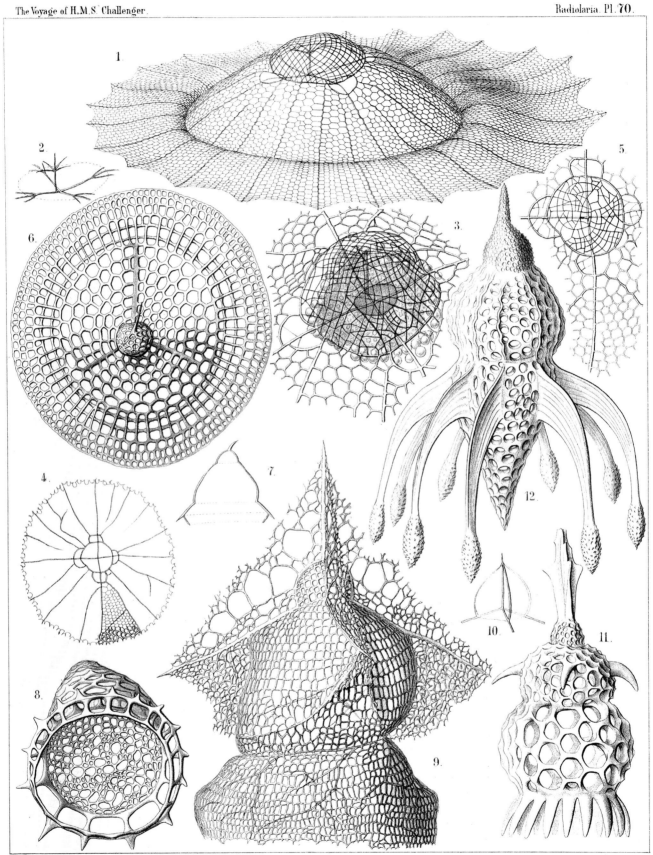

E.Haeckel and A.Giltsch,Del.

E.Giltsch,Jena, lithogr.

1-5. THEOPHORMIS, 6. THEOPILIUM, 7. 8. CLATHROSTOMIUM,
9. 10. PTEROPILIUM, 11 PTEROCODON, 12 THEOPHATNA.

The Beginning:
Water, Earth, Sun, a Complicated Story

It was 4.6 billion years ago that our planet was formed in a gaseous atmosphere out of gigantic conflagrations made up of elements, dust, boulders and meteorites. The agglomeration was molten, so the heat due to impacts was substantial. While cooling, water became liquid. An ocean was formed. Then life emerged . . . Was it born in a mud puddle as Darwin thought in the middle of the 19th century? In a pool with a flash of lightning like Miller proposed in the 1950s? Or near the volcanic hydrothermal vents in deep-sea trenches as suggested by submersible dives in the 1970s? Did it come from space in blocks of ice that cooled and grew the nascent ocean? Will we know the answer one day?

Microorganisms were already living in the water at least 3.5 billion years ago. Capable of living in environments devoid of oxygen, these microorganisms derived their energy from chemical modifications of metals and gases. This process still exists today, especially in hydrothermal hot springs.

Microbial activity thus began to transform the planet. Indeed, bacteria—photosynthesis champions—capture light energy, combine water and carbon dioxide to produce sugars and other organic molecules, and release oxygen. Thus it was in the primitive ocean that life, the food chain and carbonate rocks began.

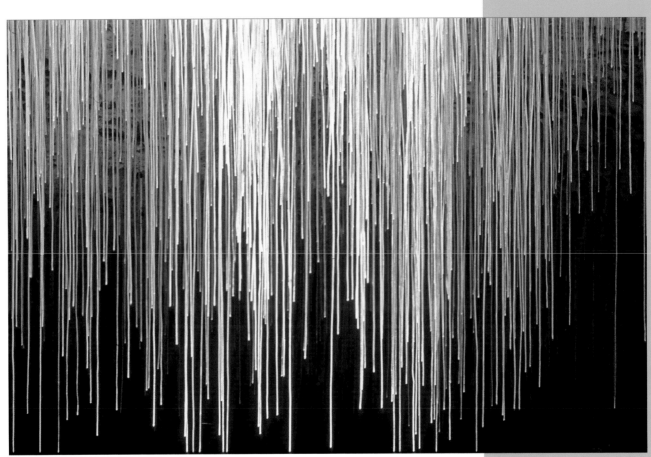

Soda straw stalactites in the Choranche Caves (Vercors, France), arising at least in part from bacterial activity. This cavity has exceptionally long straws: up to 3.2 m (approx. 10.5 ft).

If each discovery raises more questions than it answers,
then the unknown increases as our knowledge increases.

Pierre Vathomme

How Long
Has Life Existed?

There is no question that life has existed on Earth for 2.8 billion years. However, some evidence of life seems to be more than 3 billion years old. In fact, we only know about the first life forms indirectly, through the chemical modifications they introduced. The indicators left behind are in the form of rocks made from living things. They look like layered mats—hence their name stromatolites (microbial mats; see p. 62). Surprisingly, stromatolites are not organisms with mineralized shells made of calcite or silica, which are the oldest documented organisms, but rather their shells are made of organic matter. The first remains of living organisms are microscopic and in the shape of small creased purses, consisting of organic matter similar to chitin, the substance that forms insect shells. These organisms lived in the water, probably as isolated individuals. This chitinous matter is so stable that it piques the curiosity of chemists. Knowing how to manufacture a material with such stability could have many applications for industry.

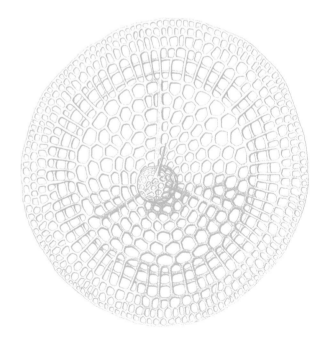

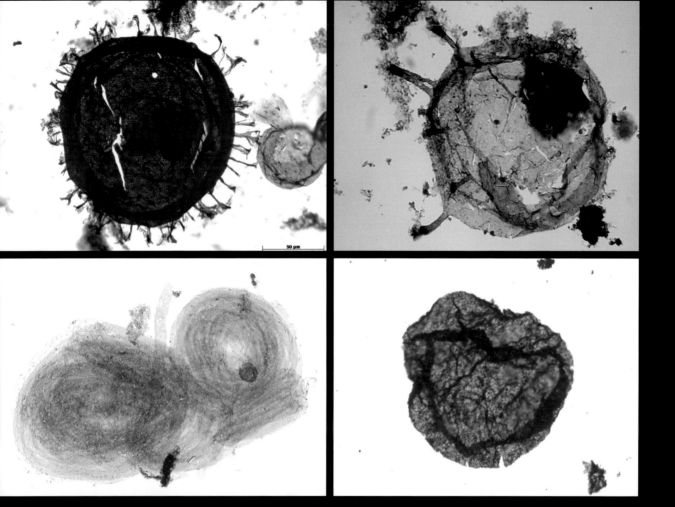

Top left: Acritarch protist (*Shuiyouspheridium macroreticulatum*). It is 1.65 billion years old. Its diameter measures approximately 0.1 mm. Provenance: Ruyang Group, China.

Bottom left: Possible cyanobacterium (*Obruchevella* sp.) dating back one billion years. Size of each element: approximately 0.24 mm. Provenance: Mbuji-Mayi Supergroup, Democratic

*Everything existing in the universe
is the fruit of chance and necessity.*

Democritus

When Did Life
Begin?

Life shares this strange question with time and consciousness. It is a familiar concept, and yet it is particularly difficult to define, perhaps because what it means remains unclear.

The organic (living) world and the mineral (inert) world have long been considered opposites. Since the 19th century, this duality has become less and less apparent. It is not an easy task to define what life is, and there is not a complete consensus. Nevertheless, some requirements for life are unanimously accepted, such as metabolism (extraction of nutrients from the environment, transformation and excretion of waste) and the ability to reproduce.

The origin of life remains a mystery. Nobody knows exactly where, when or how it came to be. If we refer to the first indirect traces of life (carbon bound to living matter), then life may have appeared around 3.8 billion years ago, but this idea remains controversial. This date is a bit of a "cap," seeing as the last meteorite impact event (which occurred between 3.9 and 4 billion years ago) was so intense that it erased all traces of older life, supposing that life existed before then. We could assume that life may have emerged before 3.8 billion years ago, but this is not verifiable. However, it is recognized as very likely that life appeared between 3.8 (end of the Late Heavy Bombardment) and 3.4 billion years ago (the age of

fossil structures in Australia that are construed, without certainty, to be stromatolites).

Only one thing is certain: life already existed 2.8 billion years ago (traces of unambiguous fossil life) and for nearly one billion years, archaea and bacteria seem to have been the only forms of life on our planet.

This rock sample shows fine colored laminae, which are layers of bacterial biofilms. These layered mats are called stromatolites (meaning mats of rocks). This sample from Bolivia was formed nearly two billion years ago (east of the Andes, south of Cochabamba, Cochabamba Department, Bolivia). Size: 38×31×2 cm.

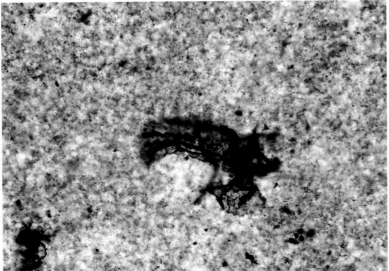

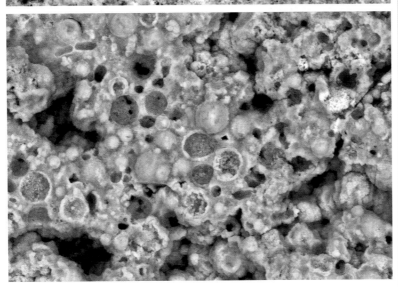

From top to bottom:
Fusiform and spherical microfossils (black) in organic matter inside a rock (quartzite) from Australia (Farrel, Pilbara) that date back three billion years. Their biological origin is supported by a range of chemical indicators. Photo taken by optical microscopy. Width approaching 0.015 mm and length approaching 0.025 mm.

Remains of a microbial mat (blackish) in rock (quartzite) from Australia (Farrel, Pilbara) dating back three billion years. Photo taken by optical microscopy. Width approaching 0.07 mm and length approaching 0.2 mm.

Bacteria in a phosphate rock. The small spheres with rounded markings are approximately 0.002 mm in size. Photos taken with a scanning electron microscope. Phosphate (coprolite) sample from Morocco from the Cretaceous Period (60 million years ago).

*Those who say, "It's obvious; you only have to look," live in an impressionistic world.
They believe they are observing the world, when they are really only observing
the impression that the world makes on them.*

Boris Cyrulnik

From Microbial Mats
to Rocks

Regarding past life forms, we are only aware of what they were able to leave behind for us. Small isolated cells without hard parts are unlikely to have survived. However, the activity of these small but numerous organisms did leave behind traces. Thus, we find rocks with laminated layers that we call stromatolites (from the Greek *stromato*, meaning "layer," and *lithos*, meaning "rock") that date back 3.5 billion years.

The first living organisms to absorb carbon dioxide (CO_2) and release oxygen (O_2) modified the chemical balance of the environment most noticeably by leaving limestone deposits. These carbonate structures are stacked layers of accumulated fine particles that were trapped by microbial mats, which are communities of microorganisms dominated by photosynthetic cyanobacteria. This limestone bears the mark of living organisms, if only by its shape: it appears in the form of layered waves or "cauliflower" concretions. Stromatolites are not living things but are rather sedimentary structures produced by bacterial biofilms or filaments. These structures sometimes have very thick layers. Stromatolites are found in rocks that are several billion years old, and modern-day stromatolites can be found, still in the process of forming, in water.

Left:
A stromatolite, cut in half, clearly shows laminae development and growth. Some structures reveal interruptions in growth and even periods of dissolution, a kind of karstic microcavity. This sample comes from the Limagne, France, and is 20 million years old (Aquitanian).

Opposite, from top to bottom:
Stromatolites have existed since the beginning of life, forming either immense planar structures or pouf-like ones, such as these living ones in Shark Bay, Australia.

Ancient stromatolites. Here "rounds" have joined together, forming a rocky bank in which these ancient structures are found. They are 450 million years old (Ordovician), Ottawa, Canada.

Oxygen,
That Poison Waste

The first organisms appeared in an atmosphere without oxygen. They used sunlight to convert water and carbon dioxide (CO_2) into sugar (carbohydrates). Therefore, they transformed fleeting light energy into a chemical compound as a way to store energy. This transformation process released a by-product, a type of waste, which was oxygen, or oxygen gas (O_2). This waste product, a strong oxidizing agent, was of course harmful to organisms of this era. Hecatombs were among the bacteria that did not tolerate oxygen. But life adapts itself to everything, and some organisms eventually integrated this toxic product into their metabolism to such an extent that it could be believed today that oxygen is a fundamental element of any life form.[2]

Precursors of phytoplankton, called cyanobacteria (photosynthetic bacteria), consumed carbon dioxide (CO_2) and produced a considerable amount of oxygen, modifying the composition of the original atmosphere, which was then rich in nitrogen, methane and carbon dioxide. As carbon dioxide is a greenhouse gas, its reduction in the atmosphere would later have an effect on the climate.

Subsequently, oxygen spread to the upper parts of the atmosphere, where it could be transformed into ozone (trioxygen, O_3), which acts like a shield against some ultraviolet radiation. Only then could life emerge from the water.

2. Organisms that have adapted to O_2 are in a better position for survival than those that have not because reactions with oxygen (aerobic) produce much more energy than those without oxygen (anaerobic).

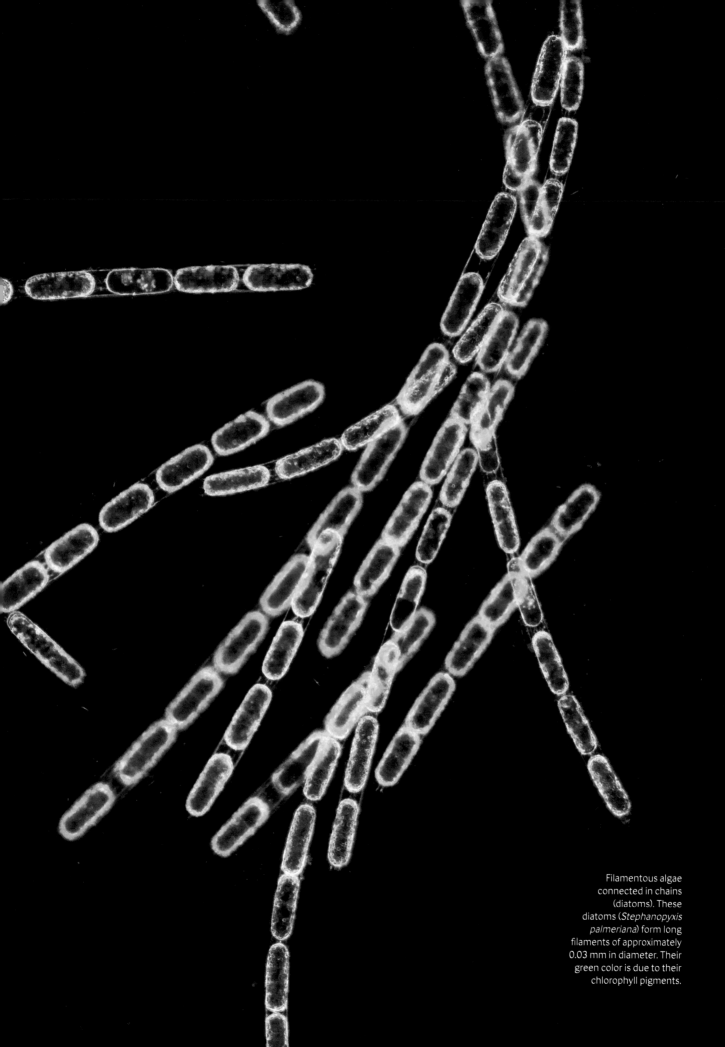

Filamentous algae
connected in chains
(diatoms). These
diatoms (*Stephanopyxis
palmeriana*) form long
filaments of approximately
0.03 mm in diameter. Their
green color is due to their
chlorophyll pigments.

In the most arid steppes, the contemplators will always be amazed.
Their naturalistic eye will reveal the most microscopic trace of life.
Their souls will be able to transcend miserable things.
Looking at a pebble, Leonardo da Vinci imagined a mountain.

Sylvain Tesson,
Petit traité sur l'immensité du monde
(Short treatise on the immensity of the world)

Minerals
Arising from Life

When the Earth was formed, some elements were concentrated in the Earth's core and others in the mantle. This process allowed a number of minerals to form. There were thus approximately 250 minerals in existence when the planetesimals agglomerated into small planets. Mercury stopped at this planetesimal stage and consists of 350 minerals. Mars, which was momentarily able to retain water on its surface, has 500 minerals, 150 of which resulted from interactions with water. On Earth, a tectonic system developed that drew rocks deep below the surface, thus offering a range of new temperature and pressure conditions that allowed minerals specific to metamorphic activity to develop. Consequently, we can identify 1,500 minerals in the oldest soils on our planet.

Over the course of its development, life has provided new physical and chemical conditions. Oxygen produced by bacteria, for instance, changed the chemical conditions on the surface of our planet. These changes allowed for new forms of minerals. With the presence of oxygen, the minerals that rose to the surface of the Earth—whether they had a base of iron, nickel, manganese, copper, uranium, or cobalt and so on—were oxidized ("rusted"). This process represents a key step in the history of the mineral world because 2,900 minerals appeared as a result. Among the most beautiful of these minerals are turquoise, azurite, malachite and so forth. In addition, living organisms themselves know how to produce certain minerals, such as mother-of-pearl. This ability has had an impact on the morphology of our planet and on how our planet functions.

The development of life's ability to give rise to minerals distinguishes Earth a bit more from other planets.

Mother-of-pearl is the iridescent inner layer of some mollusk shells. It is biosynthesized by the organism and made of aragonite crystals and conchiolin. Here, an abalone shell, also called sea ears or *Haliotis*. Hunting for mother-of-pearl for gastronomy or decoration (guitar rosettes, etc.) is one of the reasons it is becoming increasingly scarce.

Photo taken with an electron microscope of a fossilized ammonite shell (*Leymeriella* sp.) from the Cretaceous Period (Albian, approx. 110 million years ago). The shell shows layering typical of the hexagonal crystals that form mother-of-pearl.

NONE   SEI   5.0kV   X2,200   10μm   WD 11.3mm

Azurite crystals on malachite nodules (Morocco).

What is snow?
A little cold,
a lot of childhood.

Christian Bobin

Life on Earth:
Snowball Earth, the Ice of Life

While life warms the heart, it certainly cooled the climate.

Through the process of photosynthesis, trapped carbon dioxide (CO_2) led to the precipitation of calcium carbonate ($CaCO_3$). The concentration of carbon dioxide, a greenhouse gas, in the atmosphere therefore significantly decreased over time, which caused a drop in the Earth's temperature. The temperature dropped down to −50℃ (−58°F) for a few tens of thousands of years. The white surface of glaciated areas reflected more solar energy, increasing cooling and freezing. This process led to a completely frozen planet, from the poles to the equator, much like a snowball. For a few million years, the surface temperature was −10℃ (14°F).

Because most of the sun's energy was reflected, the Earth's state should have been irreversible. However, the core continued to be active. Volcanoes released gases—carbon dioxide in particular—that then accumulated because the surface of the Earth was totally frozen, meaning that mineral alteration of silicate rock (granite, gneiss, basalt, etc.) had ceased. However, this mechanism extracted a lot of CO_2 from the atmosphere, and since it was not being trapped, the presence of CO_2 increased. When CO_2 escaped into the air, through a crack for example, the greenhouse effect resumed and deglaciation could take place. This process of glaciation and deglaciation occurred several times.

Both the Earth and life-forms cause imbalances. However, they themselves—sometimes in another form—rebalance the changes they cause. Everything is a matter of equilibrium, but more specifically, a matter of dynamic equilibrium, which is to say, everything is constantly changing.

Icebergs in Greenland.

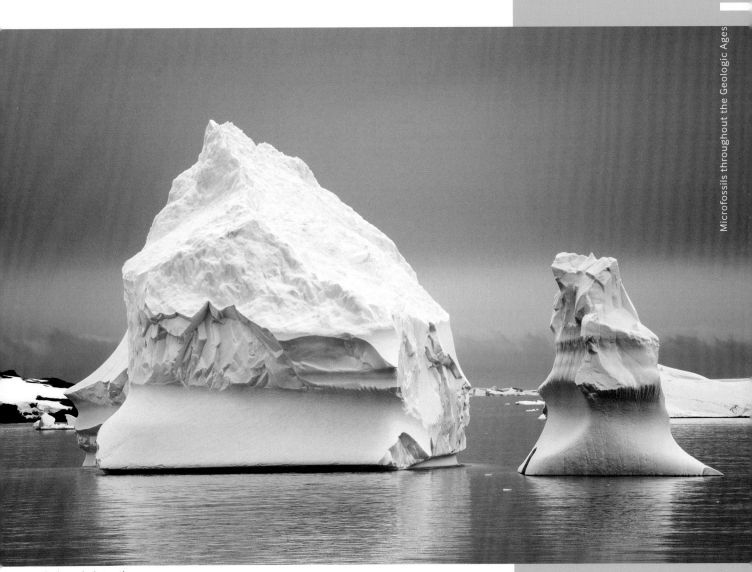

Icebergs in Antarctica.

Iron Deposits Produced by Life:
The Iron of Life

Around 2.4 billion years ago, oxygen-producing photosynthetic microorganisms changed the composition of the ocean and the atmosphere. Indeed, before photosynthetic activity, the Earth's atmosphere was dominated by carbon dioxide (CO_2) and the ocean was rich in ferrous iron (Fe^{2+}), which is green and water soluble. With the introduction of oxygen into the ocean, all of the ferrous iron oxidized and became ferric iron (Fe^{3+}), which is red and insoluble in water. Consequently, the iron precipitated and accumulated on the ocean floor. During less favorable periods, the development of microorganisms slows down and eventually stops, and with it the production of oxygen.

Therefore, today we find layers upon layers of banded iron: very thick alternating ferrous and siliceous layers, like ancient shields (found in Australia, South Africa and Brazil). More than 80% of the world's iron ore mining comes from "banded iron" deposits formed in this way. Rid of the iron that precipitated, the ocean became clearer. This allowed sunlight to penetrate deeper and photosynthesis to occur in a larger area of the water. The increased transparency of the water accelerated photosynthesis and thus the production of oxygen.

Banded Iron Formation (BIF, also called itabirite) from the state of Minas Gerais (Brazil) dating from the Paleoproterozoic Era (2.5 to 1.6 billion years ago). Millimetric to centimetric layers of iron (silver color) alternate with more siliceous layers. The layers were deposited horizontally and subsequently folded.

To learn to look away from oneself is necessary in order to see many things,
this hardiness is needed by every mountain climber.

Nietzsche,
Thus Spoke Zarathustra

Ozone Allows
Life to Leave Water

Ozone is a molecule composed of three oxygen atoms (trioxygen, O_3), whereas what we generally call oxygen is composed of two atoms (oxygen gas, O_2). Ozone is formed following a series of chemical transformations as a result of the sun's ultraviolet rays, and it accumulates naturally in the stratosphere. It began to accumulate as soon as "normal" oxygen (oxygen gas) was formed, but it was only plentiful when there was enough oxygen gas in the air.

The elements that make up living matter (proteins, DNA, RNA, etc.) are sensitive to ultraviolet rays. Ozone absorbs a significant portion of these rays, protecting living organisms. Without this ozone, life would not be possible outside of water.

The threshold amount of oxygen necessary to produce an effective ozone shield was first reached around 600 million years ago. It coincided with an explosion of species diversification. Another threshold, at which we remain today, was reached 400 million years ago. This means that enough ozone existed in the atmosphere to protect life outside of water. From then on, organisms have been able to live in emergent zones.

Radiolarian with its symbiotic algae. Here we mainly see a radiolarian (double whitish sphere) in the middle of small yellow spheres: photosynthetic algae (zooxanthellae).

Top:
In the light of day, the algae are distributed in a gelatinous mass to make the most of the available light and thus complete photosynthesis (and correspondingly absorb carbon dioxide and release oxygen).

Bottom:
In the absence of light, the algae gather around the radiolarian, most likely for an exchange function (still poorly understood).

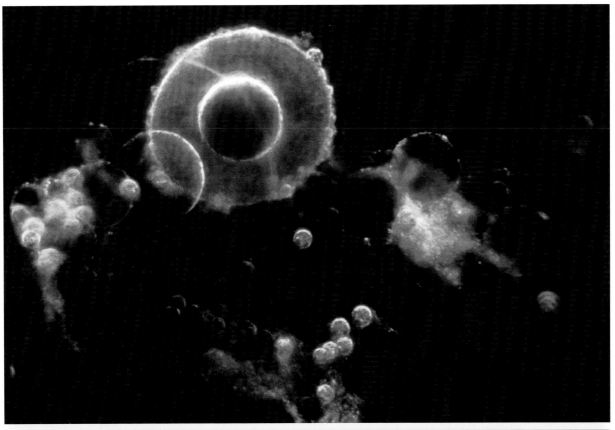

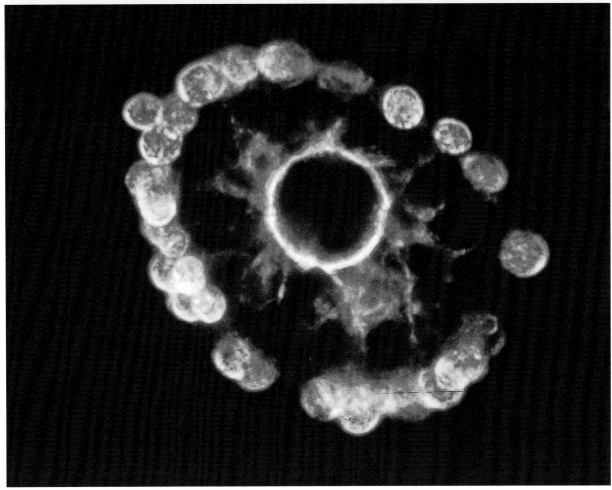

Illusion is the first appearance of truth.

Rabindranath Tagore

Most often appearances are deceiving.
We must not always judge what we see.

Molière, *Tartuffe*

The Cambrian
Explosion

Life did not develop on Earth in a steady manner. There have been periods of crisis and periods of rapid growth, both in terms of population and diversity. Warm climates have always been suitable for a wide variety of life-forms. The same is true at sea level. Life is particularly favored when rising sea levels and warm temperatures occur simultaneously.

The greatest known diversification event is called the Cambrian Explosion. It occurred 550 million years ago. At this time many organisms acquired hard (skeletal), mineralized (made of carbonates, phosphates, silica, chitin, etc.) tissues. Even today, we find more traces of snails than of slugs from the previous year in our gardens because the snails left their shells behind. For the same reason, we find traces of organisms more easily if they left mineralized skeletons. The fossil record is therefore much better documented after the development of mineralized skeletons.

We are right to wonder if what seems to us to be an explosion of life-forms in the Cambrian Period is not, in the first place, really an explosion of evidence and therefore of our awareness of them.

These phosphatic microfossils (colorized), observed under a scanning electron microscope, are fauna from the Doushantuo Formation (Guizhou, South China). They bring to mind cells in the process of dividing. Long construed to be embryos, researchers now think that they are organisms related to certain parasites of present-day fish. These cells are approximately 555 million years old (Ediacaran). The photo's width represents 0.15 mm.

Wine is sunlight held together by water.

Galileo

Plankton,
Source of Oil

Living organisms are systems in a state of thermodynamic non-equilibrium. To maintain this state, they must expend energy. On the surface of the Earth, the most readily available energy source is solar energy. Normally, this is what life uses. Thus, some organisms are capable of transforming the sun's light energy and storing it in the form of carbon molecules (such as sugars and lipids). They are called the primary producers of the food chain and include green plants and some bacteria. Other organisms feed on this form of stored energy. Further along the food chain, they themselves are consumed, and so on.

When an organism dies, its organic matter decomposes and then releases water vapor, CO_2 and stored energy. When this energy is not dissipated by oxidation upon the organism's death, it can be stored for millions of years and produce energy resources that humans exploit. This is true of some oil droplets that are preserved in the form of petroleum.

The reaction that occurs when we use fossil fuels—coal, oil or natural gas—is the reverse of the photosynthetic process that was performed a long time ago. When we burn carboniferous coal, what heats us is solar energy that was stored 300 million years ago.

Top:
Small colony of radiolarians lined up with their photosynthetic algae (yellow spheres). Here we see three radiolarians with contiguous spheres surrounded by their yellowish capsular membrane. Ten algae (zooxanthellae) surround the radiolarians. The pearly white central part is a lipid droplet, which is concentrated energy.

Bottom:
Colony of five to six radiolarians with their photosynthetic algae (small yellowish-green spheres) on the right and spicular skeletons on the left. The most visible spheres are small lipid droplets in the central capsule of each individual.

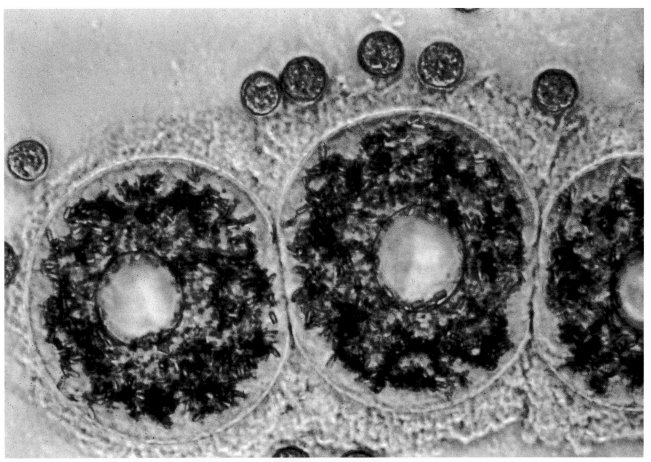

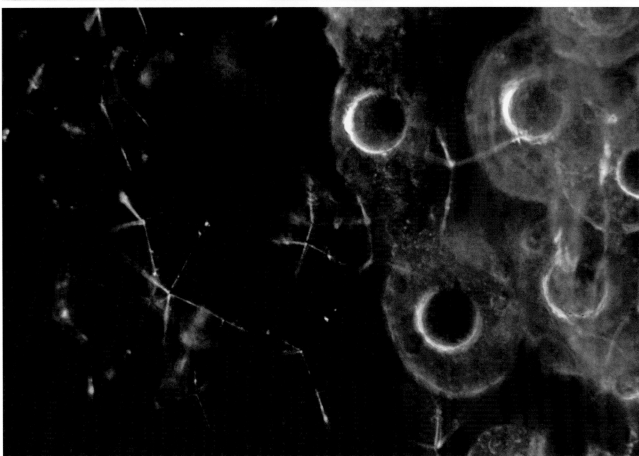

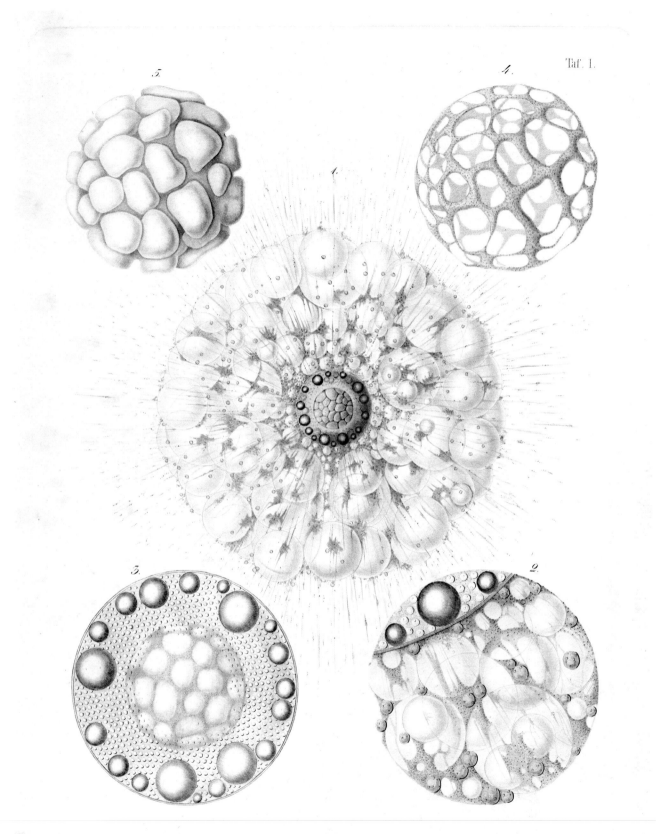

1–5. Thalassicolla pelagica, Hkl.

E. Haeckel del.

Wagenschieber sc.

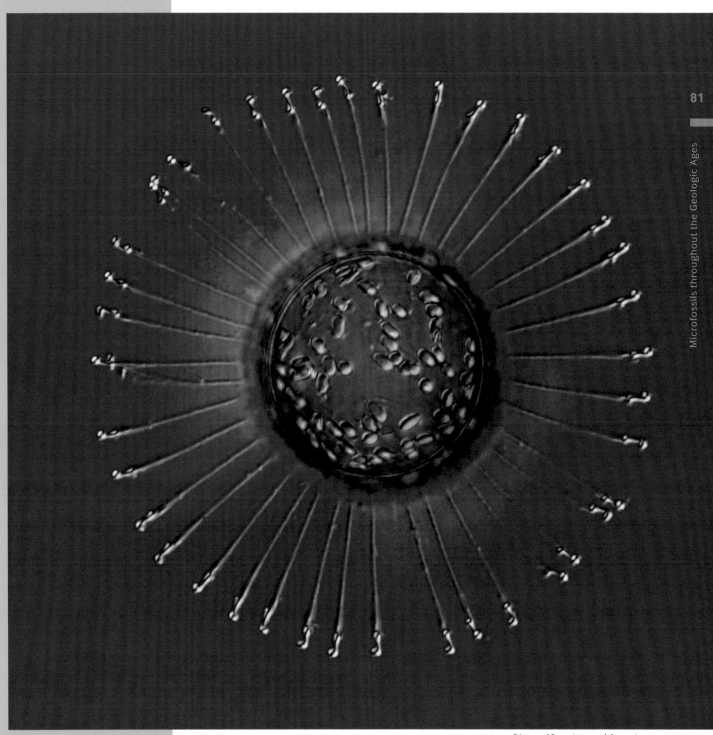

Opposite: Illustrations of radiolarians (siliceous plankton) by E. Haeckel, published in *Die Radiolarien: Eine Monographie* (Radiolaria: A Monograph), Berlin: Altas, 1862.

Diatom (*Corethron* sp.) from the Southern Ocean (Amundsen Sea). Sample provided by Eun Jin Yang. Droplets visible on the central part are lipid globules. Photo taken with an optical microscope (with Bounin's fixative). The diameter of the central part approaches 0.03 mm.

Plankton
and Mankind

Plankton is aquatic; humans live outside of water. Yet humans are closely related to plankton: every other breath of air we inhale is a gift from plankton. Bacteria and photosynthetic unicellular organisms (protists) produce as much oxygen as all terrestrial plants.

Plankton is also a large supplier of fossil fuel. The corpses and waste of planktonic organisms turn into sediment in the form of bacteria-rich flocculent particles ("marine snow"). This incessant marine snowfall has been feeding the seabed for hundreds of millions of years.

The organic part of these accumulated sediments is sometimes buried, transformed by organisms, compressed and heated. It finally produces an oily liquid, responsible for oil and gas resources. Humans draw on this carbon resource for heat, getting around, light and making a multitude of objects. At present, every year we consume the equivalent in oil of two million years of plankton buried in the ocean. In other words, in a single day we consume 11,000 years' worth of useful planktonic production.

The mineral part of the plankton accumulates in the form of ooze that, with time, becomes rock.

At any rate, as it is at the bottom of the food chain, plankton nourishes us. Without plankton, there would be no fish, and eventually, there would be less grain for chickens, and then there would not be a "chicken in every pot."

We are dependent on plankton!

Illustrations of radiolarians (siliceous plankton) by E. Haeckel, published in 1862 (*Die Radiolarien: Eine Monographie* [Radiolaria: A Monograph]).

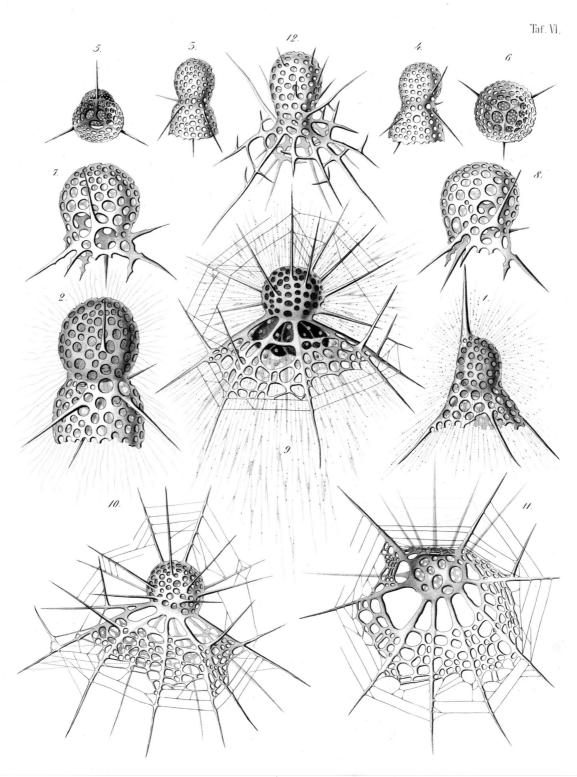

1. Dictyophimus Tripus, Hkl. 2–8. Lithomelissa Thoracites, Hkl.
9–12. Arachnocorys. 9–11. A. circumtexta, Hkl. 12. A. umbellifera, Hkl.

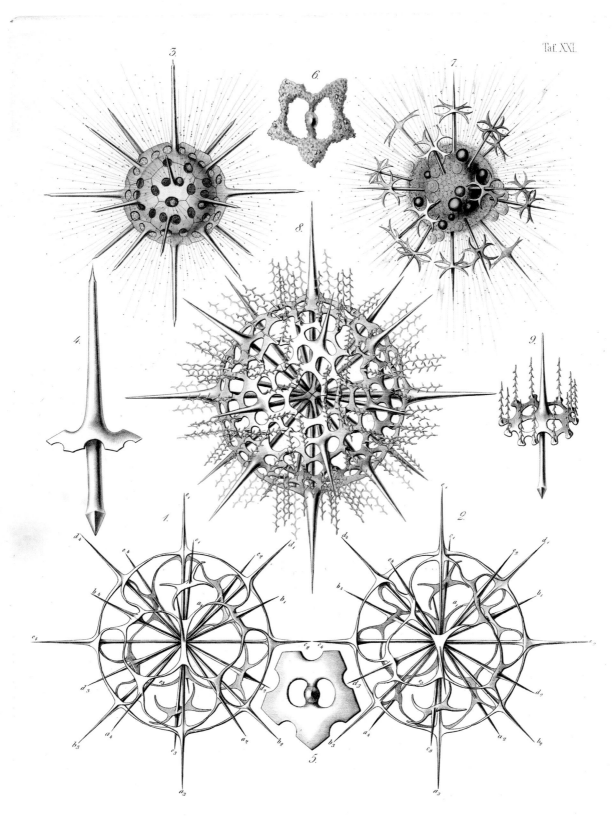

1–9. Dorataspis. 1. 2. D. bipennis, Hkl. 3–6. D. loricata, Hkl.
7–9. D. polyancistra, Hkl.

Planktonic organism: radiolarian from sediments in the Tasman Sea, which lived approximately 40 million years ago (Eocene). Photo taken with a scanning electron microscope. The largest dimension measures approximately 0.2 mm.

Change in the world is not only creation, progress, it is first and foremost decomposition, crisis.

Alain Touraine

The Microscopic World:
Susceptible to Extinction Events

Life, ever since it has been known to exist on Earth, has been constantly changing. Sometimes the changes have been violent and have affected many groups living around the world over short periods of time. On a geologic scale, these episodes are actually millions of years and are called extinction events. It is difficult to identify the events that affected the living world before mineralized skeletons existed because there are so few traces, if any at all. For organisms with a mineralized skeleton, five major extinction events have been identified, as well as many other smaller ones, that have affected both macroscopic and microscopic organisms. Depending on what caused the event, not all life-forms were impacted in the same way.

Large extinctions deeply affected living things. Just as the mammals benefited from the dinosaurs' disappearance, which enabled them to grow and diversify, some surviving species of plankton began to occupy new ecological niches following extinction events. With each event, the planktonic tree of life lost branches, but the species that survived diversified. Thus, species of centric diatoms that gradually appeared 200 million years ago survived the massive extinction of the Late Cretaceous Period (about 66 million years ago). Other species of diatoms, the pennates, appeared. Since then, centric and pennate diatoms have been particularly abundant in the cold, silica-rich waters of the polar regions.

The major extinction events, far from being curses, have been opportunities for life-forms to diversify and proliferate.

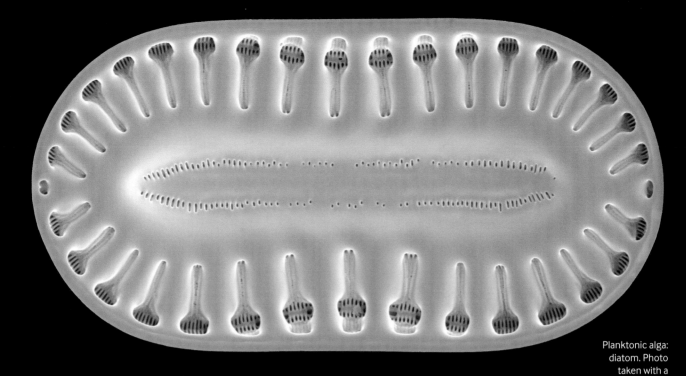

Planktonic alga: diatom. Photo taken with a scanning electron microscope. The largest dimension measures approximately 0.1 mm.

Planktonic organism: radiolarian from rocks in Turkey (Elbistan). This organism lived about 220 million years ago (Carnian). Photo taken with a scanning electron microscope. The largest dimension measures approximately 0.4 mm.

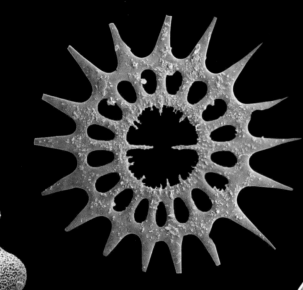

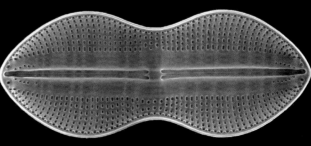

Planktonic organism: radiolarian from rocks in Turkey (Elbistan). This organism lived about 220 million years ago (Carnian). Photo taken with a scanning electron microscope. The largest dimension measures approximately 0.3 mm.

Planktonic alga: diatom. Photos taken with a scanning electron microscope. Images of the same diatom, superior view and lateral view. The largest dimension measures approximately 0.1 mm.

*Nature is a sublime book
that we must not
tire of reading.*

Camille Flammarion

The Astronomical Cycles
of the Earth

Sedimentary rocks are witnesses of the past environment and, therefore, reflect the major changes of former climates.

The sun manages the essential functions on the planet's surface. Through the energy that it provides to the Earth, the sun is responsible for the winds and the ocean currents, as well as plant growth. The sun is therefore where all food chains start. Most organisms depend on it.

The Earth rotates, but not along a perfectly fixed axis. It also revolves around the sun, but not in a perfect circle. Rather, it moves in an elliptical path that changes shape over time. All these astronomical irregularities mean that the seasons change according to nested cycles of 20,000, 40,000 and 100,000 years. Seasons influence our environment, and sed-

iments record these changes. The seafloor is the receptacle for all that falls from the water mass above. Accordingly, clays and remains of organisms that lived in the water column are found there.

When environmental conditions are favorable, organisms multiply, and a lot of remains accumulate. In less favorable years, deposits are less substantial. So depending on the year, or sets of a few thousand years, the amount of sediment that accumulates varies. Over time, the sedimentary ooze is transformed into rock with banded layers that reflect the major seasonal cycles.

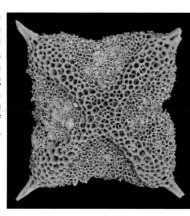

Radiolarian from Romania from the Cretaceous Period (80 million years ago). Photo taken with a scanning electron microscope. Its largest dimension is in the range of half a millimeter.

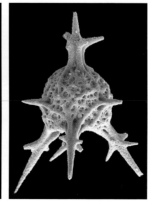

Radiolarian (Archocyrtidae) from the Triassic Period (approx. 242 million years ago). Photo taken with a scanning electron microscope. Its width is one-third of a millimeter.

Stratified siliceous rocks (radiolarites). The succession of bands (2 to 4 cm each) reflects changes in the climate (every 20,000 years). Outcrop in the forest of Keihoku-cho Katanami, Ukyo-ku, near Kyoto, Japan. Late Triassic (225 million years ago).

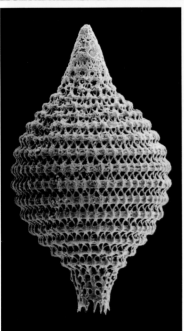

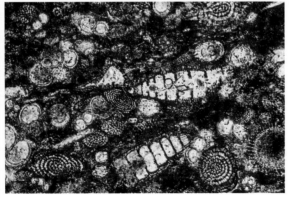

Very thin strip of rock (0.03 mm) that, held up to light, allows the microfossils within it to be seen; here are radiolarians that are approx. 75 million years old (Campanian). Sample taken during a deep sea drilling project in the Central Pacific. Their size is between 0.1 and 0.2 mm.

Radiolarian (*Mirifusus dianae*) from the Jurassic Period (150 million years ago). Photo taken with a scanning electron microscope. Its width is one-third of a millimeter.

*To explain a wisp of straw,
we must disassemble the entire universe.*

Remy de Gourmont

Expending Energy
to Make Skeletons

Organisms must spend a lot of energy collecting the necessary chemical elements for secreting their skeletons because there is not a sufficient concentration of these elements in water.

The activity of some living organisms can lead to substantial limestone deposits. This is the case, for example, of certain photosensitive bacteria that deposit large mats of calcium carbonate called stromatolites. In a way, this mineral deposit is a by-product of the activity of living things. In contrast, other organisms precipitate minerals because they need a protective covering, like shells that act as sheaths. Other minuscule organisms make skeletons that serve as both protection and support for the movements of their cytoplasm, which is a kind of colloidal substance in which the cell's organelles are found. The cytoplasm and organelles together constitute a cell. The spines of some organisms, for instance, are used as a "launchpad" for the cytoplasm to stretch and extend outward to form long filaments in search of prey.

To build its skeleton, an organism takes the chemical elements it needs from water. If an element is practically nonexistent in the water, then this process requires a lot of energy. This is the case for organisms that have a silica skeleton (diatoms, silicoflagellates, radiolarians, ebridians, etc.). Therefore, these organisms can only develop if they have a lot of food available. For example, thanks to deep water upwelling, today we can find rocks formed by the accumulation of siliceous microfossils where the ocean was once rich in nutrients.

Bloodstone (radiolarite). Greek intaglio representing the head of Athena Parthenos. Modern mount. National Roman Museum, Palazzo Massimo alle Terme.

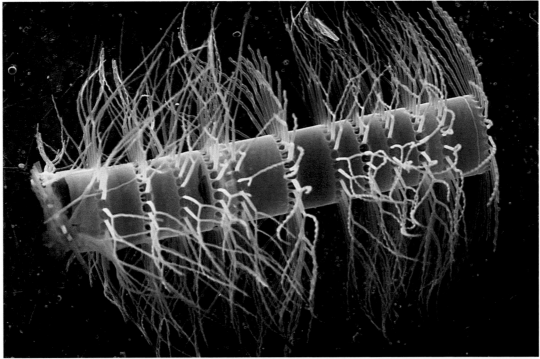

Photosynthetic siliceous alga: diatom. Long radial outgrowths emerge at the junctions between segments. Photo taken with a scanning electron microscope.

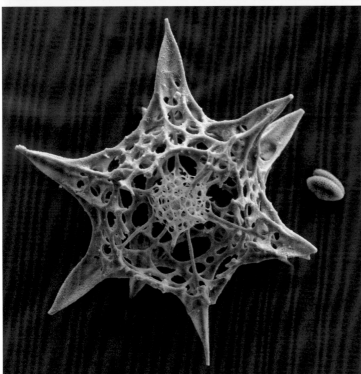

Planktonic organism: radiolarian from rocks in Turkey formed 220 million years ago. (*Triassospongosphaera latispinosa*). Photo taken with a scanning electron microscope. The largest dimension measures approximately 0.2 mm.

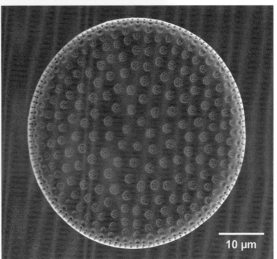

Photosynthetic siliceous algae: polar marine diatoms. Photo taken with a scanning electron microscope. Each of the two diatoms reaches 0.1 mm; the two disks are separated by 0.04 mm at most.

A Plant
Base

The food chain shows that each being, each link, feeds at the expense of another and serves as food for the next link, all the way to the end of the chain. Despite its protective skeleton, plankton is used as food for many other organisms, especially copepods, which are small millimetric crustaceans (see p. 116).

What is at the bottom of the food chain? They are beings that do not feed on other organisms but, more poetically, on sun and fresh water.

Plants in general, including plant plankton, or phytoplankton, convert solar energy into glucose, which is organic matter, and release oxygen. Diatoms, for example, which are small, unicellular algae found in seawater as well as in lakes, ponds and even wells, are very diversified. There are thousands of species. To ensure photosynthesis, they need sun. Therefore, they live in the upper part of the water column where they are exposed to the sun's rays, which is at roughly 30 to 50 meters.

Diatoms are particularly abundant in cold waters (Arctic and Antarctic). From the silica dissolved in the water, they make stiff silica envelopes called frustules.

Top:
Photosynthetic siliceous alga: diatom. Detail of the junction between two segments. Photo taken with a scanning electron microscope.

Bottom:
Photosynthetic siliceous alga: diatom (*Paralia sulcata*). Detail of the junction between a few segments showing the wall's complex perforations. Photo taken with a scanning electron microscope.

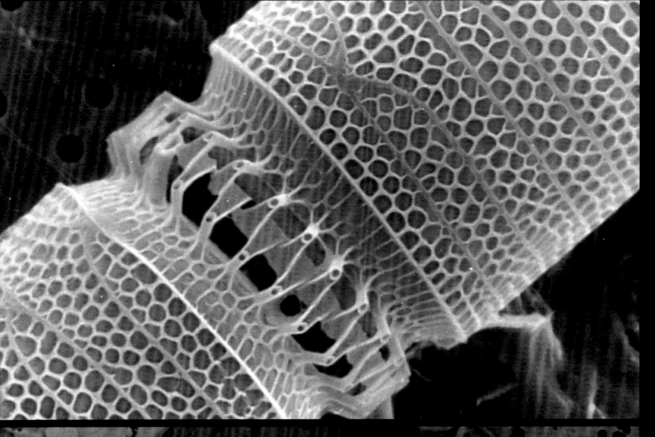

The sun, with all those planets revolving around it and dependent on it, can still ripen a bunch of grapes as if it had nothing else in the universe to do.

Galileo

Living
Together

Life seems to have existed for three billion years in the form of simple, isolated organisms. Quite quickly in terms of the geologic scale, cells began to live together. At first they did so in groups of unitary individuals without specialized functions (such as sponges). Next, certain organisms began to have specialized functions, until finally true societies of what are called multicellular organisms were formed. Multicellularity appeared at least 25 times over the course of evolution through different mechanisms, probably because of the selective advantages it affords organisms, such as the possibility to be larger or for different cells to specialize. Did this happen about 2.1 billion years ago as some suggest, or later? To this day, this way of life persists and most organisms, from protozoans (single cells) to metazoans (individual cells consisting of specialized cells), still exist in colonies. The fact that life-forms that organized into societies of cells have existed for a long time and also among very different groups of living things indicates that life in society must bring more advantages than disadvantages.

Some of these protists are occasionally associated in colonies of thousands of individuals in the same gel-like substance (cytoplasm), with no specialized cells. These colonies then grow to large volumes or into long chains of several meters (some radiolarian colonies reach 5 m [approx. 16.4 ft]).

Radiolarian colony (*Aulacantha scolymantha*) caught off the coast of Villefranche-sur-Mer, France.

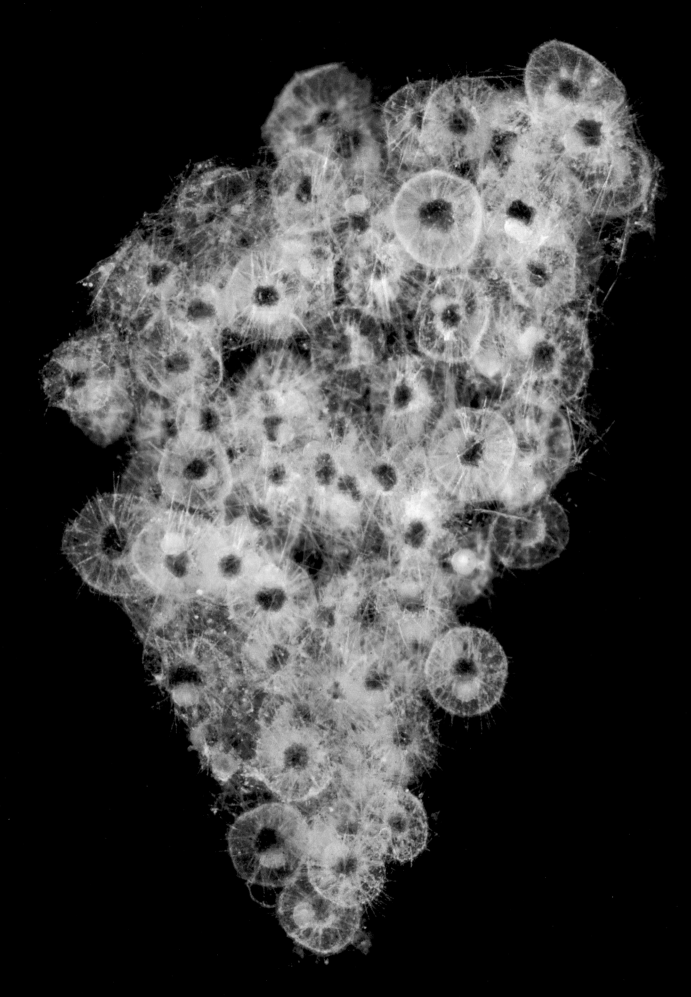

*We must learn to live together as brothers
or perish together as fools.*

Martin Luther King, Jr.

Plankton:
The Beginning of the Chain

The ocean, lakes and rivers are home to a number of microscopic organisms. These beings are not quite animals or plants as we usually know them. They are unicellular organisms called protists. The biological organization of some protists tends toward animality (zooplankton), while others are closer to plants (phytoplankton).

Many organisms feed on plankton, from the very smallest organisms to the 15-meter-long basking shark (approx. 49.2 ft) or 20-meter whale shark (approx. 65.6 ft), as well as sardines, anchovies and so forth.

The passage between each link in the food chain is accompanied by a loss of energy related to the predators' metabolism. It is estimated that for a kilogram of apex predator (such as tuna), a mass of at least 10 metric tons of phytoplankton was consumed at the beginning of the food chain.

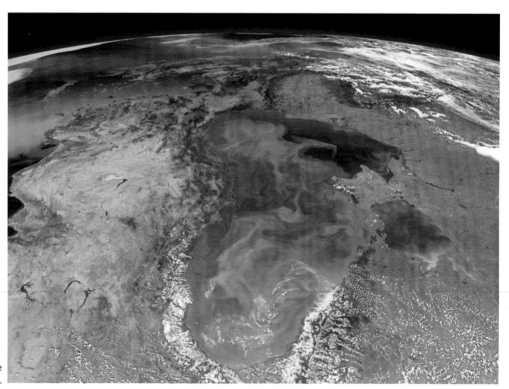

Blooms in the
Black Sea.

Bloom in the Bering
Sea near the Pribilof
Islands. Image taken
from the Landsat8
satellite.

Victor Hugo
Visionary

It is a cloud, in fact, that forms all this first depth of the ocean above the unknown bottom, and from this cloud falls, into the second depth, a rain. What rain? A rain of living things. A rain of animalcules. Here the mystery is revealed. The immensity of the microscopic is unmasked. The trembling of creation seizes you.

One could say that it is with the infinitely small that the enormity of the sea begins.

The sea has its product, the foraminifer; the ocean secretes infusoria. The molecule and the cell, these two limits of microscopic vision, so abstruse that the animal cell is not distinct from the plant cell, this Calpe and this Abyla of the infinitely small, engender, in combining with all the obscure forces suspended in the ocean, an imperceptible being. What does this being do? It built the continents under water.

The function of this atom is to replace, at any given time, the Europes, the Asias, the Africas and the Americas, which you have at this moment under your feet.

It is the exceptional worker doing extraordinary work.

There where underwater life seems to end, it is born. It asks the bottom of the monstrous cloud for waves, and, constantly, at every minute, day and night, there fall innumerable, immense eternal rains.

Dizzying analogies! It's snowing on the top of the mountains, it's raining on the ocean floor. Only that which is snowing at the top of the mountains is death; what is raining at the bottom of the sea is life.

It took the most powerful microscopes known for this molecule to take shape. Subjected to vigorous magnification, the atom was revealed, the infinitesimal confessed, and we saw appear under the lens a shell, frail, fine, transparent, as white as snow and as pure as crystal.

This shell is the cave of the infusorian; this shell is the workshop of the foraminifer.

An almost incomprehensible thing—this shell, thinner than the most fragile glass, is always whole. This shell is miraculously virginal. Never a break, never a crack, never an erosion; its edges are sharp, its points are piercing. The whole sea weighs peacefully on this fragility.

The law of the seafloor is known today. It is the shell of the foraminifer who uttered it.

One observes the foraminifer; one can only note the polycystines.

And probably, and certainly, one could say, as the foraminifer is a giant for the polycystine, there is further below another being for whom the polycystine is a colossus, and so on, oh terror! Until infinity is exhausted.

Such is the law, and it trembles there, and one loses oneself there; moths of moths, lice of lice, scabies of mites, vermin of vermin, abyss of the abyss.

The foraminifer secretes lime; the polycystine is made of silica. The vital cosmic fluid builds the Pyrenees with this lime and the Andes with this silica.

It is there that, among the sponge spicules and a few rare diatomaceous beings that are simultaneously plants and animals, crawls the worm of this sepulchre, the foraminifer, having under it another worm, the polycystine; and it is there that, lugubriously lit by the quantity of light that can pass through a window pane that is twelve thousand meters thick, in silence, in immutability, in solitude, the atom works in the world.

There are more stars in the sky than infusoria in the sea. In the sky the polycystine is called the sun.

—Prose philosophique, première partie
  (Philosophical Writings, Part One)

Small glass lacework
In your sheets of clays and oxides
You will cross spaces
Without ever aging.

Jean-Yves Reynaud

A Terrible Purgatory
for the Beginning of Eternity

Usually, when a planktonic organism dies, its organic matter disintegrates and disappears. Its mineral skeleton is then subjected to the aggression of the seawater, which dissolves it in a few days. It is a true work of art that disappears. But it happens that Mother Nature shows some compassion toward her own creations. Although for certain forms to traverse time, she has condemned them to be devoured! The copepods (see "Copepods," p. 116), for example, which are arguably the most common planktonic organisms, snatch other plankton for food. The organic matter is digested, whereas the mineral matter of the indigestible carbonate or siliceous skeletons is discharged in fecal pellets, which are small oblong or spherical droppings. What a terrible end! Not at all. This apocalypse is not an ending, but a passage that will allow this waste to persist for thousands or even millions of years to come. Indeed, this muddy waste will supplement the sediment on the bottom of the seas. Composed of clay and organic matter, it constitutes a protective mold that shelters the skeletons inside. These true gems are thus protected from the assault of time. They can be found in a fossilized state millions of years later.

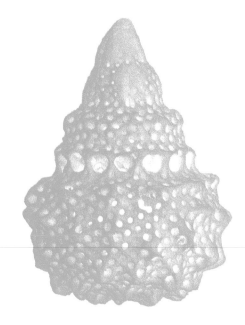

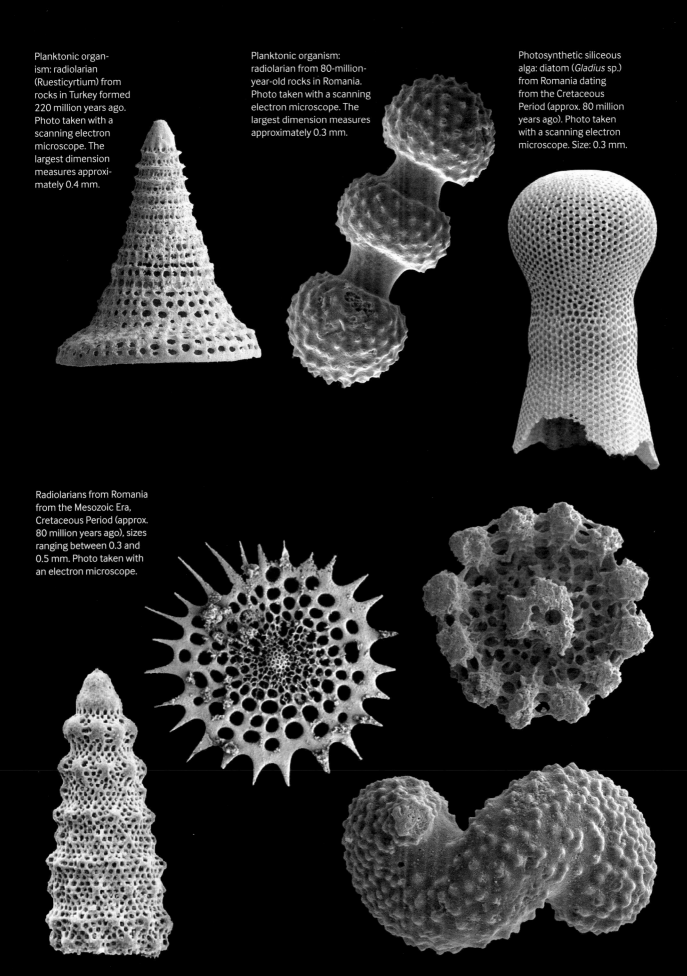

Planktonic organism: radiolarian (Ruesticyrtium) from rocks in Turkey formed 220 million years ago. Photo taken with a scanning electron microscope. The largest dimension measures approximately 0.4 mm.

Planktonic organism: radiolarian from 80-million-year-old rocks in Romania. Photo taken with a scanning electron microscope. The largest dimension measures approximately 0.3 mm.

Photosynthetic siliceous alga: diatom (*Gladius* sp.) from Romania dating from the Cretaceous Period (approx. 80 million years ago). Photo taken with a scanning electron microscope. Size: 0.3 mm.

Radiolarians from Romania from the Mesozoic Era, Cretaceous Period (approx. 80 million years ago), sizes ranging between 0.3 and 0.5 mm. Photo taken with an electron microscope.

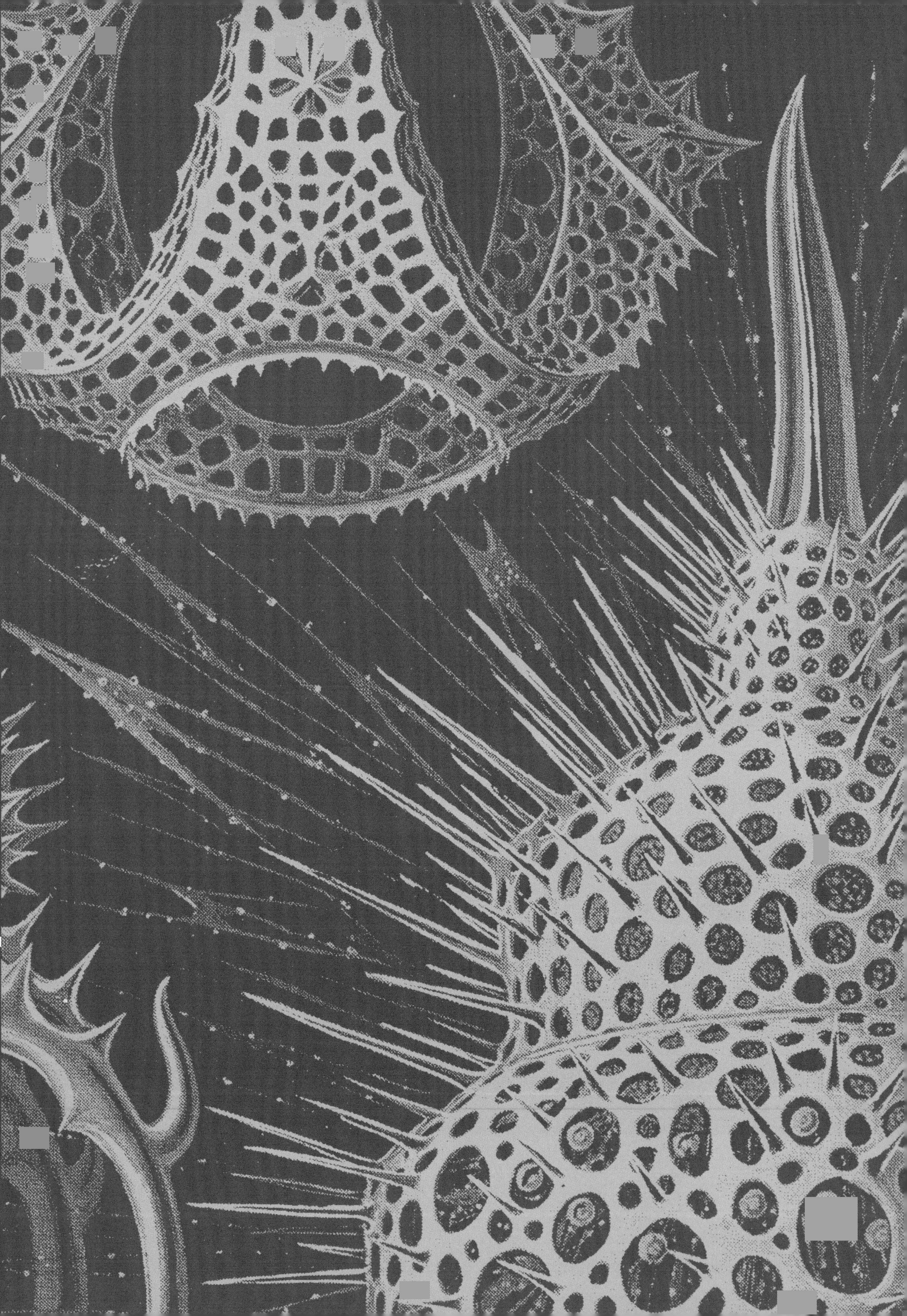

The Diversity
of Microfossils

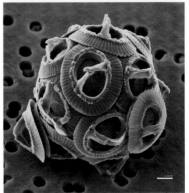

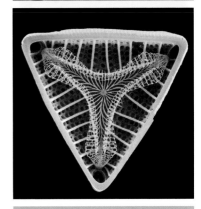

Classifying
Microorganisms

In the 17th century, Swedish naturalist Carl von Linné laid the foundation for classifying the living world. Ever since, paleontologists have not stopped discovering and describing species and classifying them according to their visible morphologies. Similar species were grouped together into the same genus, similar genera into the same family and so on.

For living organisms, genetics is increasingly considered when making and verifying groups and is used to recognize common ancestors. For the fossil world, there is not sufficient genetic data; their genes are not preserved. Only morphological data from their skeletons is available. Therefore, microfossils are broadly classified using the morphology of living organisms. This method is largely justified by the fact that there is usually a strong connection between an organism's skeletal morphology and its biology.

Organisms have long been classified according to their differences; today, it is done based on their similarities.

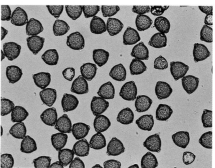

Microscopic elements in organic matter (cucumber spores) that could become microfossils. Photo taken by optical microscopy.

Top left:
Calcareous nannoalga in present-day oceans: coccolith (*Gephyrocapsa oceanica*). Photo taken by scanning electron microscopy. The largest dimension approaches 0.006 mm.

Top right:
Microcrustacean (ostracod). Photo taken by scanning electron microscopy. The largest dimension approaches 0.9 mm.

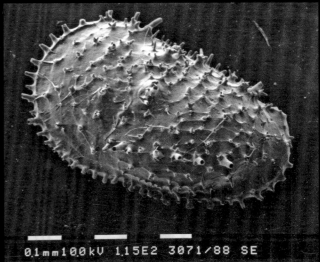

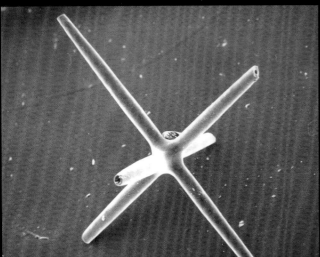

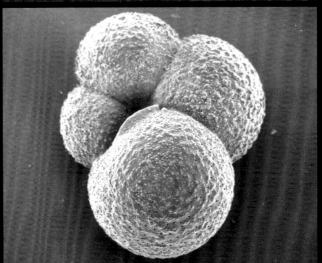

Bottom left:
Siliceous microorganism (sponge spicule). Photo taken by scanning electron microscopy. The largest dimension approaches 1 mm.

Bottom right:
Calcareous microorganism in present-day oceans (*Neogloboquadrina pachyderma*). Photo taken by scanning electron microscopy. The largest dimension approaches 0.3 mm.

Little boxes on the hillside
Little boxes made of ticky-tacky
Little boxes, little boxes
Little boxes all the same.

Malvina Reynolds and Pete Seeger

Finely Perforated
Plant Boxes

Diatoms are known to have existed since the Jurassic Period, 200 million years ago, but they were not abundant until the Cretaceous Period (about 150 million years ago). Nowadays they alone generate a quarter of our planet's oxygen.

Diatoms are microscopic unicellular algae (0.01 to 0.15 mm) with siliceous shells (valves) that generally look like a round Camembert cheese box—a lid that fits on top of a slightly smaller bottom. Their name comes from the Greek *diatomos*, meaning "cut in half." The internal wall of each valve has a great variety of structures, whose fine, consistent features make some useful as resolution test targets for optical microscopes. Some diatoms are banded together (see "Expending Energy," p. 90, and "A Plant Base," p. 92), but most of them live in isolation or in motion, or sometimes they are sessile. They are found in any body of water that is exposed to light.

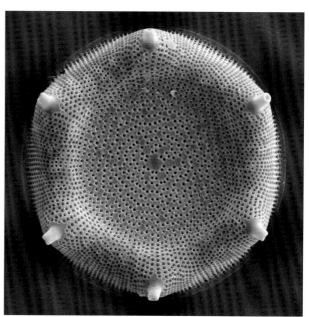

Photosynthetic siliceous alga: diatom. Photo taken with a scanning electron microscope.

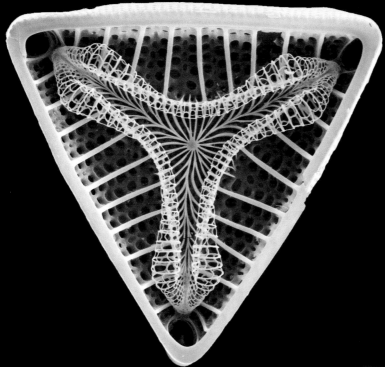

Photosynthetic siliceous alga: diatom (*Entogonia* cf. *formosa*) from sediments in Romania (Păușești-Otăsău, Getic Depression) dating back 12 million years (Middle Miocene, Tertiary). Photos taken with a scanning electron microscope. The largest dimension is in the range of 0.1 mm. External and internal views and interior detail. The outer shell is finely perforated and the interior reveals more complex structures.

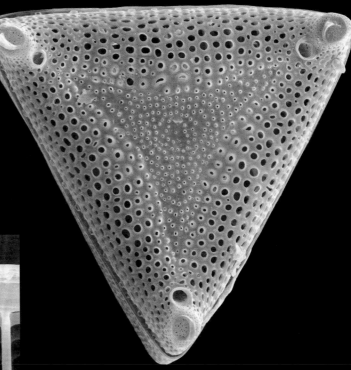

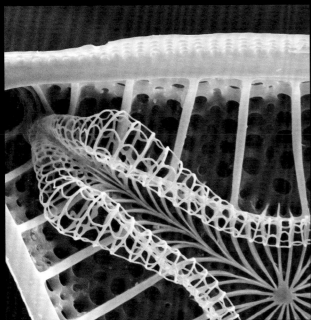

Microfossils
of Various Sizes and Behaviors

Microfossils are fossilized organisms of a small size (less than 1 mm) that require suitable tools, such as microscopes, to be observed. Almost invisible to the naked eye, they have nothing in common with one another except for their size, and in no case are they the same size as the organisms they once were. Indeed, what is the relationship between the nummulite *Nummulites millecaput*, a unicellular organism that reaches nearly 10 centimeters (almost 4 in) in diameter while its congeners are submillimetric, and pollen grains that are a few micrometers across in spite of the fact that they come from a tree that is several meters high? We also come across extremely small organisms measuring only a few micrometers that, when they join forces, are likely to feed on other organisms that are several centimeters in size. For instance, colonial radiolarians feed on the larvae of amphipods (small crustaceans).

These organisms are all very different. Each has its own specific function and role to play in the planktonic ecosystem. No one organism is more important than another, in the same way that it would be unwise to say that a car's steering wheel is less important than its engine. In the symphony of nature, the orchestra is made up of very different elements, and each one plays its part.

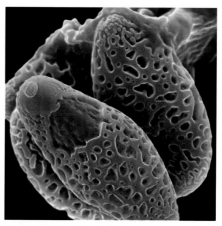

Spores of present-day bolete mushrooms (*Austroboletus mutabilis*). Photo, colorized, taken by scanning electron microscopy. The largest dimension approaches 0.005 mm.

sponge. Here is a single spicule (*Septrintus richardi*) of sediment from Australia dating from the Eocene (40 million years ago). Photo taken by scanning electron microscopy. The largest dimension approaches 0.1 mm.

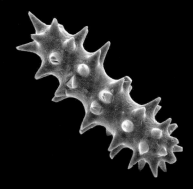

Siliceous plankton, radiolarian (*Svinitsium depressum*) from rocks in Sardinia dating from the Cretaceous Period (Valanginian, approx. 135 million years ago). Size at the largest dimension: approx. 0.2 mm.

Calcareous microfossil: foraminifer (*Elphidium* sp.). Photo taken by scanning electron microscopy. The largest dimension approaches 0.5 mm.

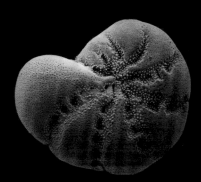

Radiolarian diet: predation. Colonial radiolarians feeding on young amphipods, the few orange-colored masses, two of which have visible eyes. The small yellow dots are symbiotic algae that live in the radiolarian colony. Only their rounded shape can be seen. Photo taken by optical microscopy.

Phytoplankton,
Planktonic Plant Life

Phytoplankton refers to all photosynthetic planktonic organisms. It encompasses cells without nuclei (photosynthetic bacteria and cyanobacteria) as much as unicellular organisms with nuclei (protists such as diatoms, dinoflagellates and coccolithophores). These two groups have in common that they draw their energy from light through photosynthesis. Protists contain specialized organelles called chloroplasts (small green organelles from which their name is derived, from the Greek *chloros,* meaning "green"). Membranes perform photosynthesis as well. Organic matter is synthesized from carbon dioxide and water. It is generally accepted that half of all oxygen is produced by phytoplankton and the other half is provided by land plants.

Phytoplankton also plays an important role in atmospheric carbon sequestration and consequently in climate regulation. Bacteria and protists represent more than 90% of the ocean's biomass. Various microorganisms produce molecules (dimethyl sulfide, or DMS) that play an important role in cloud and climate formation. Green and golden algae, dinoflagellates, coccolithophores and *Phaeocystis* (foam left behind by the tide) are all examples of such microorganisms (see "Plankton Manages the Climate," p. 196).

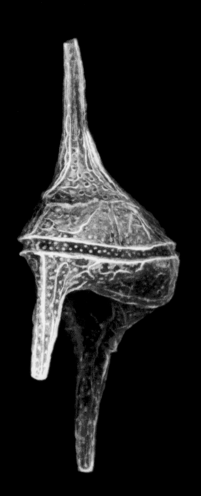

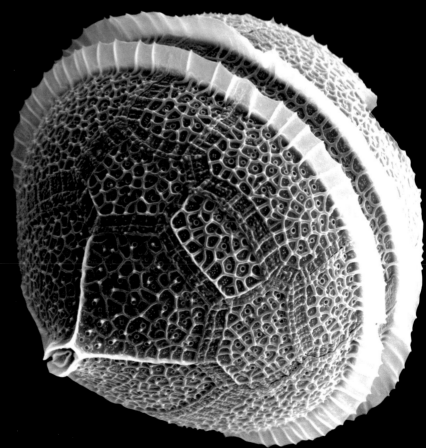

Top:
Dinoflagellate (*Ceratium candelabrum*). The theca (sheath) has been made visible by a fluorescent blue-green molecule, and the chloroplasts appear in red. Photo taken by confocal optical microscopy.

Bottom:
Most dinoflagellates, such as *Protoperidinium* sp., shown here, build envelopes made of cellulose plates. These plates vary in their ornate appearance. Photo taken by scanning electron microscopy. Size: approx. 0.03 mm.

Mercenaries with a Whip:
Dinoflagellates

Dinoflagellates are phytoplanktonic protists that measure between 0.02 and 0.15 mm. They have existed for at least 500 million years (since the Cambrian Period). Unlike other protists, dinoflagellates move very quickly, thanks to their two flagella. They are usually autotrophs because they are photosynthetic, but some species also feed on bacteria and other protists. Others still are parasitic. Therefore, they present characteristics of both plants and animals.

Dinoflagellates have organic armor that usually consists of very ornate cellulose plates. Despite their armor, they are food for many other organisms, especially for copepods, which are small millimetric crustaceans (see "Copepods," p. 116).

When the organism dies, only its armor made of organic matter is likely to be preserved. Their remains have been used effectively to date soil and also to restore environments and climates.

Dinoflagellates sometimes proliferate into veritable plankton blooms. These blooms are monitored because they are sometimes responsible for toxins that poison marine life, including aquaculture.

Protozoans in organic matter: dinoflagellates. Photo taken by electron microscopy. Size: approx. 0.2 mm.

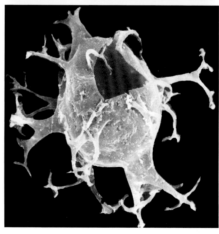

Protozoans in organic matter: dinoflagellates (*Achomoramulifer* sp.). Photo taken by electron microscopy. Size: approx. 0.2 mm.

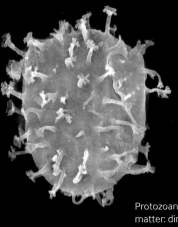

Protozoans in organic
matter: dinoflagellates
(*Kiokansium williamsii*).
Photo taken by
electron microscopy.
Size: approx. 0.2 mm.

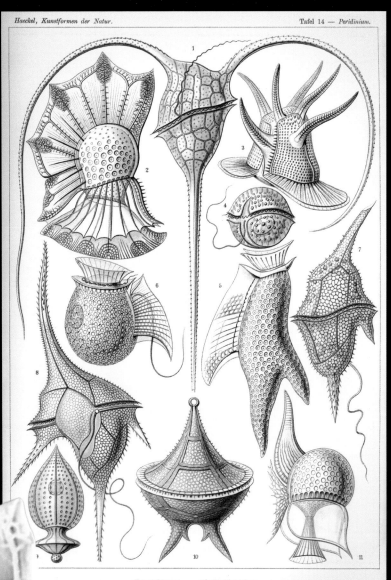

Protozoans in organic
matter: illustrations of
peridinians published by
E. Haeckel, Plate 14, in the
book *Kunstformen der
Natur* (*Art Forms in Nature*),
1899–1904.

Protozoans in organic
matter: dinoflagellate.
Photo taken by
electron microscopy.
Size: approx. 0.2 mm.

Wide-Open Spaces Intoxicate:
Ebridians

Ebridians are planktonic marine protozoans that have a small internal silica skeleton. Their name comes from how they move: probably because they have two flagella in the back, their movement seems disorderly rather than methodical like a military parade, so they evoke the movement of an inebriated person (*ebrius* means "intoxicated").

Ebridians feed on other protozoans, especially small algae such as diatoms. Their biological status is still difficult to establish. Sometimes they are closer to algae, like dinoflagellates or silicoflagellates, or to animals, like radiolarians. But their skeletons also call to mind certain spines (made of actin) seen in sponges. As ebridians are rather scarce, they are seldom studied and therefore remain rather poorly understood, even though they have existed for 100 million years (since the Cretaceous Period).

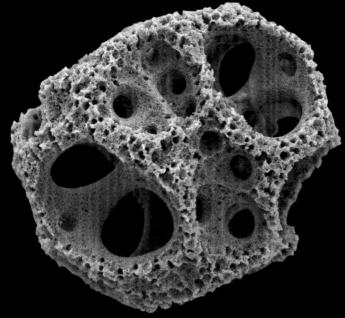

a) Skeleton. *Ammodochium speciosum* from Russia from the Eocene (about 35 to 55 million years ago). Their largest dimension reaches 0.04 mm. Photo taken by electron microscopy.

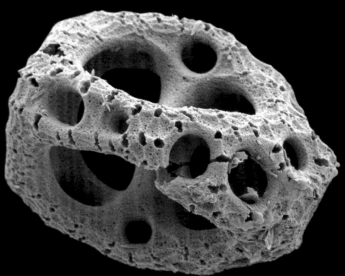

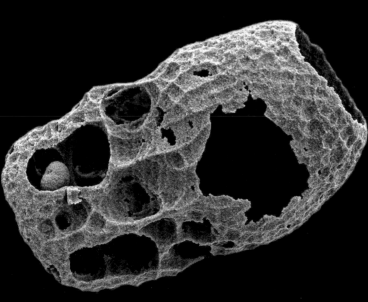

b) Skeleton. *Ammodochium* sp. from drilling projects (site 748B) on the Kerguelen Plateau (southern Indian Ocean) dating from the Middle Eocene, approx. 40 million years ago. The double lorica skeleton reaches 0.06 mm.

The most beautiful experience we can have is the mysterious. It is the fundamental emotion that stands at the cradle of true art and true science.

Albert Einstein

Copepods

Copepods are small crustaceans that live in all types of water, whether salt or fresh. The term copepod comes from *kope*, meaning "oar," and *podos*, meaning "foot," in reference to their feet, in the shape of oars, which they use to move. Small in size, between 0.2 and 10 millimeters, they are, in terms of number of individual organisms, the most abundant multicellular organisms on the planet (between 60% and 80% of the zooplankton biomass).

Copepods have a single median eye called a "naupliar eye" and do not have gills or a carapace, which means they are found very rarely in a fossilized state.

Copepods are known to have existed since the Cretaceous Period (nearly 100 million years ago), but some have also been found in sediments from the Carboniferous Period (300 million years ago). Molecular biological data suggests that they have existed since the Cambrian Period (500 million years ago).

Copepods feed on other microscopic planktonic organisms: dinoflagellates, diatoms, radiolarians and so on. The mineral skeletons of their prey are then discharged with their fecal waste. These small droppings sometimes constitute the majority of the sedimentation and represent a chance for the skeleton of these planktonic prey to be preserved in a fossilized state (see "A Terrible Purgatory," p. 100, and "White Cliffs of Manure," p. 184).

Copepods are eaten by jellyfish, shrimp, fish, whales and so forth. Every year, copepods produce 40 billion metric tons of protein, far more than the 0.26 billion metric tons of farm-raised meat produced worldwide!

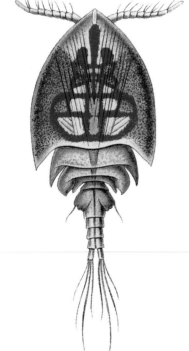

Plankton in organic matter: illustrations of present-day crustaceans (Copepods) published by E. Haeckel, in the book *Kunstformen der Natur* (*Art Forms in Nature*), 1899–1904. Size: between 0.3 and 2 mm.

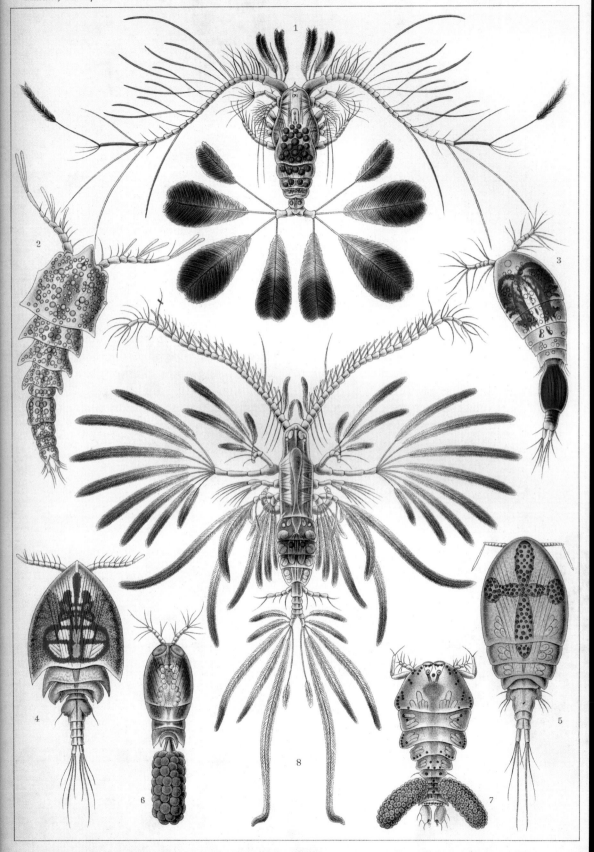

Copepoda. — Ruderkrebse.

*Butterflies are just flying flowers on a holiday
when nature was overflowing
with creativity and fertility.*

George Sand,
Contes d'une Grand-Mère
(*Tales of a Grandmother*)

Pteropods:
Microscopic Hermes

Pteropods are mollusks that have a wing-shaped appendage on each side of their body that they use for swimming, hence their name, meaning a foot in the form of wings (*ptèros*, meaning "wing," and *pode*, meaning "foot"). Yet having a winged foot seems to be the only thing they have in common with Hermes, the Greek god of travelers. For that matter, they are also sometimes called butterfly gastropods, sea butterflies or even sea angels. Unlike most gastropods that move by crawling on their stomach (hence their name, *gastero*, meaning "stomach," and *pode*, "foot") and only whose larval form can swim in the water column, pteropods are able to swim in open water. Their size ranges from less than a millimeter to a few millimeters, so they can be classified as microfossils or macrofossils. Their shells are made of aragonite, which is a rather unstable mineral under normal temperature and pressure conditions, so they can be easily dissolved. This is all the more true because many pteropods have such thin shells that they are transparent. With the least bit of dissolving, the shell disappears and the chances of it being preserved in a fossilized state are, to say the least, low.

Pteropods are known to have existed for nearly 65 million years (throughout the Tertiary).

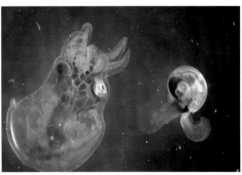

Small octopus (*Octopus* sp.) on the left, and a pteropod.

Pteropod (*Limacina rangii*) in the Southern Ocean. Doesn't it look like a flying snail?

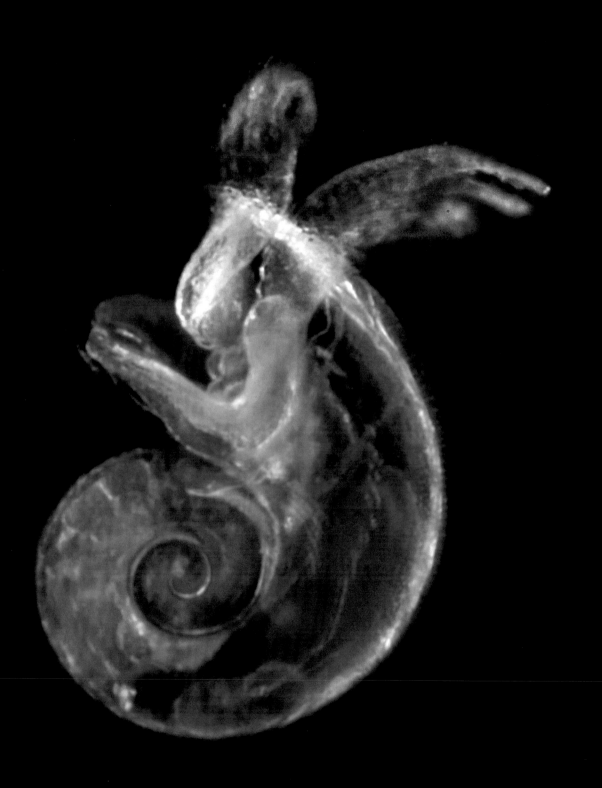

Pteropod (*Limacina* sp.) in water. A kind of transparent snail that has wings to fly in water.

Still Unknown:
Acritarchs

These organic-walled microfossils appear to be unicellular organisms. We do not know much more about them than that. Indeed, their biological affinity remains uncertain, as their name suggests; "acritarch" comes from the Greek *ákritos*, meaning "uncertain" or "confused" ("without critic"), and *archae*, meaning "archaic." As a result, it is almost certain that disparate organisms are grouped together under the global term *acritarch*. Therefore, their classification is artificial; it is merely practical.

Acritarchs look like vesicles. Their size is often between 0.02 and 0.1 millimeters, and their highly variable shapes range from spheres to cubes.

What are they? Many acritarchs are probably resting cysts of green algae, others are probably dinoflagellate cysts (see p. 112), others still are possibly egg cases of small metazoans and undoubtedly others may belong to other groups.

Acritarch remains are known to have existed for nearly 3.2 billion years (Middle Archean, before the Proterozoic). In other words, they are the oldest known fossils. They became especially abundant at the beginning of the Paleozoic Era (about 500 million years ago). They declined rapidly and disappeared almost completely thereafter. The vast majority of acritarchs are found in river sediments and/or associated with aquatic organisms. They are very useful for dating ancient soil and determining the paleoenvironment of a locality.

120

Acritarchs from Morocco from the Ordovician Period (450 million years ago). Photos taken with a light microscope. Overall sizes between 0.03 and 0.05 mm.

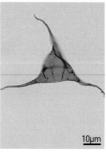
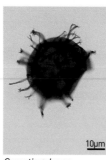

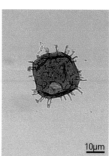

Veryhachium trispinosum group.

Cymatiogalea sp.

Frankea breviuscula.

Coryphidium sp.

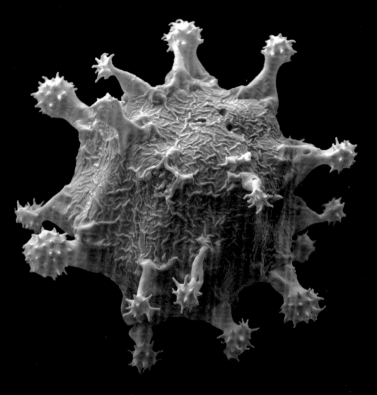

Acritarch from the Paleozoic Era. *Visbysphaera* sp. from Gotland (Sweden) from the Silurian Period (430 million years ago). Size: approx. 0.05 mm. Image taken with a scanning electron microscope.

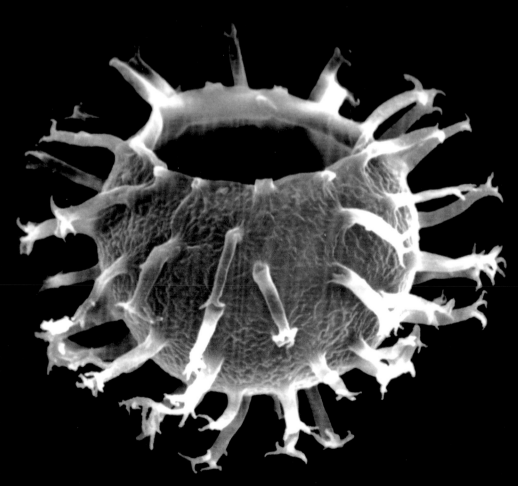

Acritarch from the Paleozoic Era. *Stelliferidium* sp. from the Algerian Sahara from the Ordovician Period (480 million years ago). Size: approx. 0.05 mm. Image taken with a scanning electron microscope.

Anyone who is no longer capable of wonderment is a dead man.

Albert Einstein

Whips
for Forward Motion

A host of cells are ciliated. In 1677, a Dutch student by the name of Ham reportedly told Antoni van Leeuwenhoek (1632–1723), who developed the microscope, that he had seen animalcule-like organisms in sperm that were moving with a tail (a flagellum). These were among the first living elements observed under the microscope. Many other microscopic organisms were discovered thereafter.

Cilia and flagella have the same microtubular structure. They sense and transmit signals from their environment, creating movement and currents that are essential to the motion and nutrition of the cells. Among planktonic organisms, ciliates are protists that possess motile cilia on their surface. There are currently about 10,000 species. Their size varies between 0.1 and 0.01 millimeters.

Until 2007, the oldest fossils known dated from the Ordovician Period (450 million years ago). Even older fossils were then found in the Doushantuo Formation in Guizhou Province in southern China. They date back 580 million years (to the Ediacaran Period).

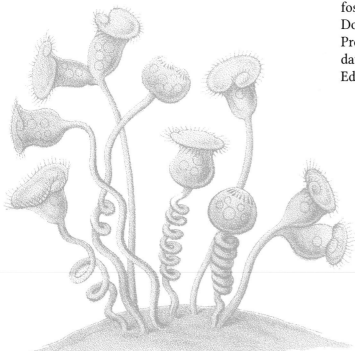

Illustrations published by
E. Haeckel, Plate 3, group
of ciliates, in his book
Kunstformen der Natur
(*Art Forms in Nature*),
1899–1904.

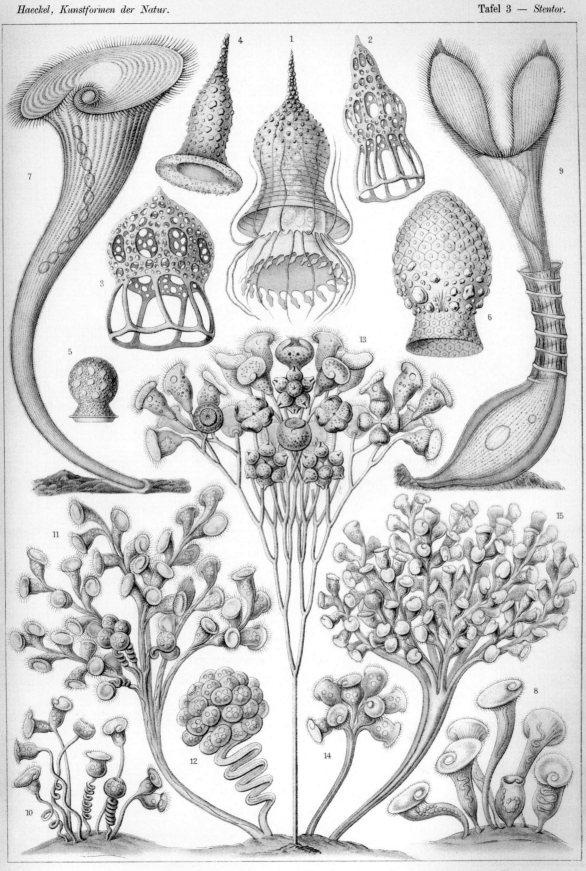

Ciliata. — Wimperlinge.

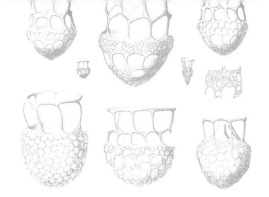

You love life, now you've only to try to do
the second half and you are saved.

Dostoevsky,
The Brothers Karamazov

Small Mesozoic
Vases

Among the ciliates are tintinnids. They include several hundred living species. While they can be found in fresh water, they are found in abundant quantities mainly in marine waters in the range of a few hundred to a few thousand individuals per liter. They have a covering called a lorica, which is a name borrowed from Latin because its form is similar to the armor breastplates, called loricae, worn by Roman soldiers. The lorica is formed from a framework of proteins. Generally they are shaped like an amphora, vase, corselet or trumpet, and some are decorated with sticky particles.

These protists are attached to the bottom end of a lorica. At different times, they either take refuge there or emerge. In the latter case, they then put out their cilia, which create currents to drive prey to their mouths. They feed on small algae and bacteria. Links in a food chain themselves, tintinnids are then the prey of copepods (tiny crustaceans) and fish larvae.

Among the most common ciliate fossils, Calpionellids (from the Greek *kalpion*, meaning "small vase") are known to have existed starting from the Jurassic Period (150 million years ago). They have a calcareous skeleton. They did not become abundant until the Creta-ceous Period (about 140 to 150 million years ago). A group similar to present-day tintinnids, they were extremely abundant during the Cretaceous Period and are very useful in dating rocks.

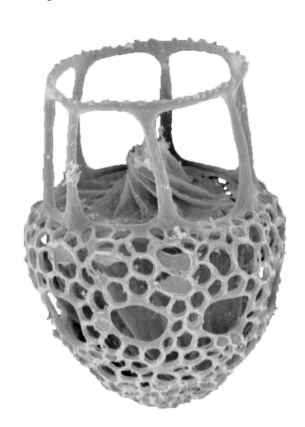

Dictyocysta grandis from Yemen. Their size reaches 0.05 mm in length. Photo taken with a scanning electron microscope.

Tintinnid specimen
(*Codonellopsis orthoceras*)
with various skeletons of
algae stuck on its shell (a
dinoflagellate and different
coccoliths: *Syracosphaera
pulchra, Caliciosolenia* sp.,
*Rhabdosphaera claviger,
Calcidiscus leptoporus,
Thoracosphaera heimi,
Heliocosphaera carteri* and
Scyphosphaera apsteini).
This specimen was caught
in the bay of Villefranche-sur-
Mer, France. Its maximum
width is a little less than
0.1 mm. Photo taken by
scanning electron micros-
copy.

*For a sieve, it is not
a fault to have holes.*

Lebanese Proverb

Foraminifers
Have Holes!

Among the microscopic organisms that populate, or populated, the seas, there is a group distinguished by its perforations. Its members are thus called foraminifers (from the Latin *foramen*, meaning "hole," and *-fer*, meaning "bearing").

Abundant for many hundreds of millions of years, foraminifers are important because they form one of the most abundant and diverse groups of fossils. Many foraminifers have a skeleton, usually mineral, called a test (from the Latin *testa*, meaning "rounded bowl," which gave us the French word *tête*, meaning "head"). They secrete their mineral skeletons, which are sometimes made of organic matter or silica but are most often made of carbonate, or

else their skeletons are made only of agglutinated sediment particles, especially if the species lives in the sediment. Foraminifers grow by building new chambers in their test. These chambers have a geometric arrangement specific to each species. They can be rectilinear, curved, coiled or even cyclic and can also be uniserial or multiserial. These arrangements can also be found in combination and can be even more complex.

Pseudopods emerge from apertures, allowing the animal to interact with its environment through sensing, feeding and moving. With these cytoplasmic extensions, foraminifers wrap around and absorb all kinds of prey, from bacteria to mollusks, as well as algae, larvae and various waste products.

Calcareous protozoans:
illustrations of different
foraminifers published
by E. Haeckel in his book
Kunstformen der Natur
(*Art Forms in Nature*),
1899–1904.

Tiny crowns of silica spines
What Lilliputian and drowned Christs lost you
Getting off their nebulous planktonic perch?

Jean-Yves Reynaud

Siliceous and Flagellated:
Silicoflagellates

Silicoflagellates are both autotrophic (photosynthetic) and heterotrophic (predators of prey) planktonic marine organisms. Their photosynthetic activity forces them to live only in the upper part of the water column that is exposed to light. Their siliceous internal skeletons consist of a network of bars, sometimes resembling radiolarian skeletons but much simpler. They represent only between 1 and 2% of the siliceous components in marine sediments. Nevertheless, as silicoflagellates are widely distributed in the oceans, they are valuable when other organisms are absent. Like many other organisms, they are most abundant in food-rich zones where cold, nutrient-rich currents move up from the depths.

Even within the same species, their morphological details vary a lot according to the specimen, which makes it difficult to use silicoflagellates to date rocks. This disadvantage is offset by the fact that their structure is very sensitive to the environment. Therefore, they are potentially good markers for paleogeography.

Appearing in the Early Cretaceous Period (around 110 million years ago), silicoflagellates became common in all oceans in the Late Cretaceous Period (around 80 million years ago) and became most numerous in the Miocene (around 10 million years ago).

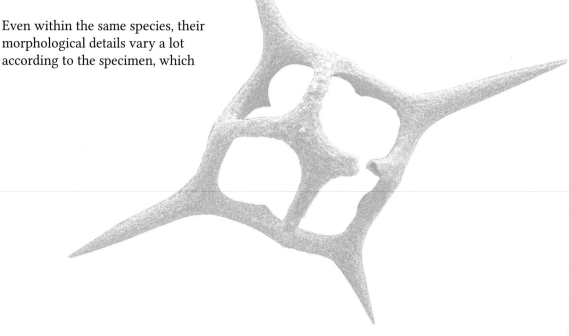

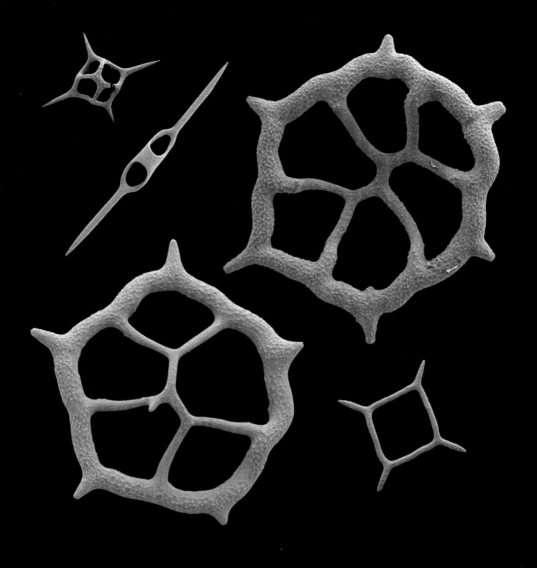

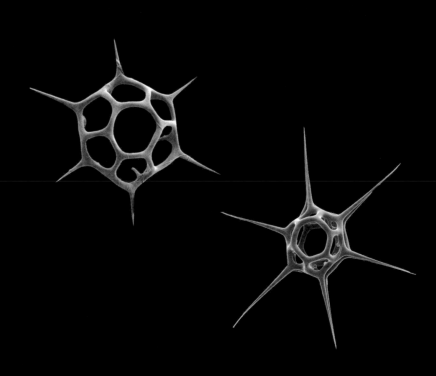

Top:
Silicoflagellate skeletons from drilling projects (site 748B) on the Kerguelen Plateau (southern Indian Ocean) dating from the Middle Eocene (approx. 40 million years ago). Their size varies between 0.05 and 0.12 mm. Photo taken with a scanning electron microscope.

Bottom:
Two silicoflagellate skeletons (*Dictyocha speculum*) from the Southern Ocean. Size: approx. 0.02 mm. Photo taken with a scanning electron microscope.

The work of nature is much more difficult
to understand than the book of a poet.

Leonardo da Vinci

Ostracods:
Animal Micro-Pistachios

Most microscopic organisms are unicellular. Multicellular ones are scarce. Ostracods are among the latter group. They are small cousins of crabs, or rather hermit crabs, who find shelter in small shells with two valves that evoke pistachio shells (their name comes from the Greek word *ostrakon*, meaning "shell, carapace," which also gave us the word "ostreiculturist").

Their carapace, most often calcareous, is articulated in the dorsal portion with a hinge. Their average size is between 0.15 and 1 to 2 millimeters, but some species reach 30 millimeters. Only the extremities of some appendages emerge ventrally from the carapace when these animals move on the substrate or when they swim. Their eggs are sometimes so resistant to desiccation and to the gastric juices from the animals that eat them that they can be ingested and then dispersed over considerable distances, for instance by birds.

Ostracods occupy all marine and freshwater environments and water sources with deep sediments. They can therefore be used as indicators of current or past environments (paleoenvironments). Like foraminifers (see p. 126), they have also been used as bioindicators to detect the presence of certain pollutants, for example in Slack Bay in France (Pas-de-Calais) or in Hiroshima Bay in Japan. Ostracods have existed for more than 450 million years (since the Ordovician Period) and have been evolving ever since. Therefore, they are effective tools for dating rocks. Approximately 7,000 current species are known.

Shells of microcrustaceans (ostracods) from Madagascar that are approximately 100 million years old (Albian-Cenomanian). They are millimetric in size.

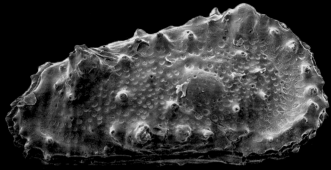

Venericythere dictyon (van den Bold, 1966)

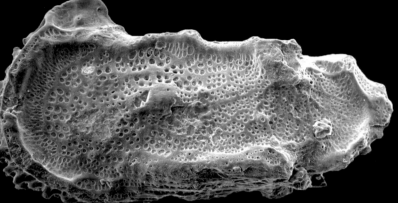

Ponticocythereis spinosa (Whatley & Titterton, 1981)

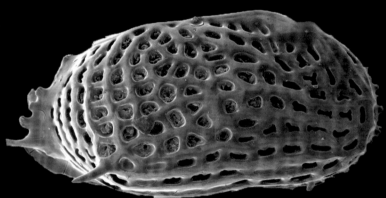

Venericythere dictyon (van den Bold, 1966)

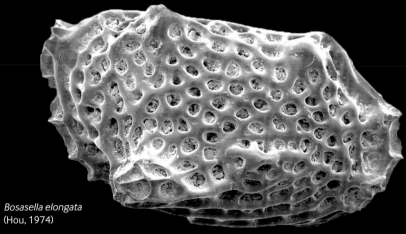

Bosasella elongata
(Hou, 1974)

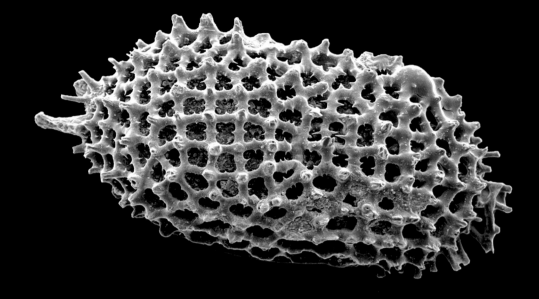

Pistocythereis cribriformis (Brady, 1860)

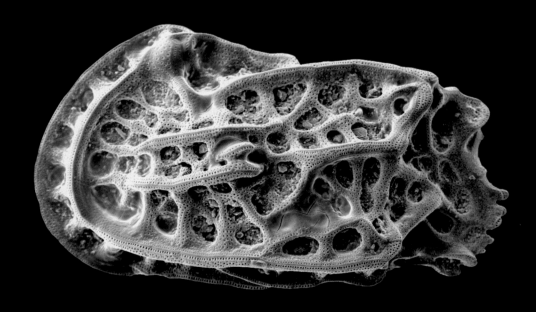

Bosasella elongata (Hou, 1974)

Ponticocythereis ichtyoderma (Brady, 1890)

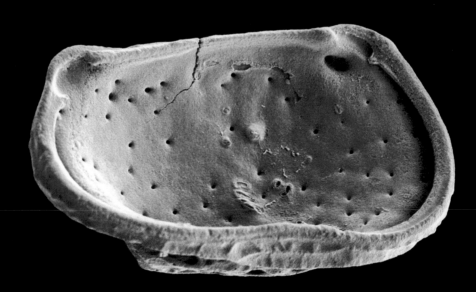

Internal view of a shell.

> *If the sea does not overflow
> it is because Providence,
> in its wisdom,
> has placed sponges there.*
>
> Alphonse Allais

Sponges

Sponges, long thought to belong to the plant kingdom, are mostly marine aquatic animals that are usually attached to the substrate. They are found at all depths. They are considered to be metazoans, but in fact, they are colonies of unspecialized cells with a rudimentary organization. Their nervous system is very primitive and diffuse. They do not have a mouth, anus or any other differentiated organ (excretory, respiratory, etc.). We could therefore say that sponges are "barely" metazoans in the same way that we think of certain colonial radiolarians as being "almost no longer" protozoans.

Incidentally, there is a word used for this: syncytium, which means "cells together" (*syn-*, meaning "together," and *-cytium*, meaning "of cells"). It underlines the ambiguity of not really being one or the other. Sponges are known to have existed for at least 600 million years (since the Ediacaran Period).

Sponge bodies are "nonliving" masses that are home to small, living cells. Their interstitial skeletal elements (spicules) are what make them rigid. These microscopic spicules are made of calcium carbonate, chitin or silica. The spicules are what are found in a fossilized state.

Illustrations of hexactinellid sponges and some of their spicules by E. Haeckel, Plate 35, from his book *Kunstformen der Natur* (*Art Forms in Nature*), 1899–1904.

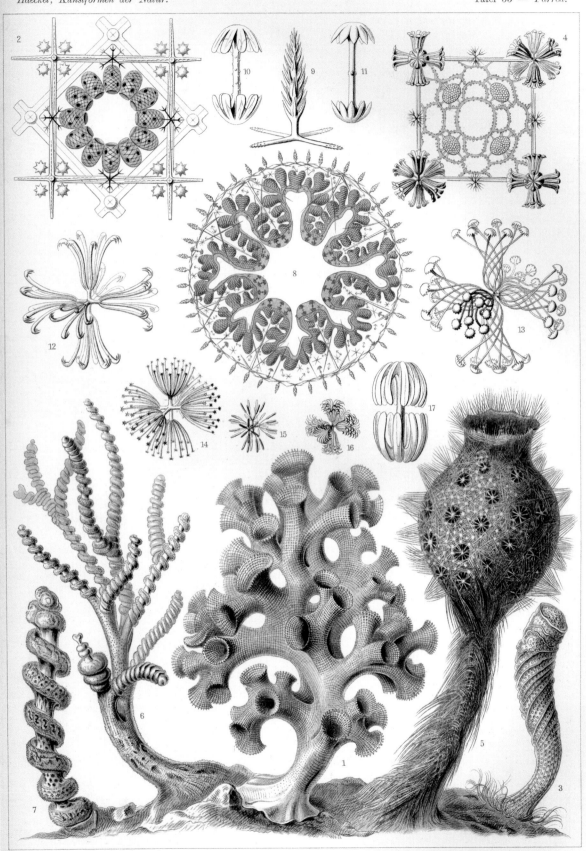

Hexactinellae. — Glasschwämme.

Amphidisc spicule of an hexactinellid sponge from a deep drilling site in the Southwest Pacific Ocean that dates from the Middle Eocene (40 million years ago). Photo taken with an electron microscope.

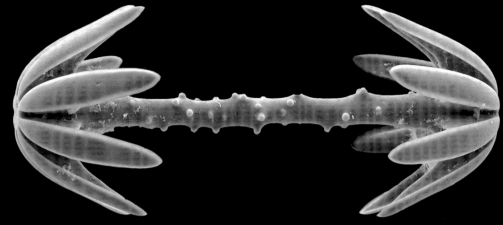

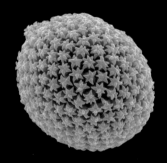

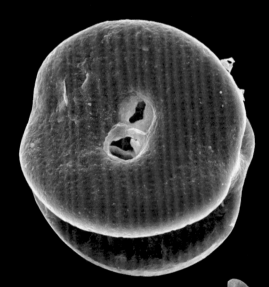

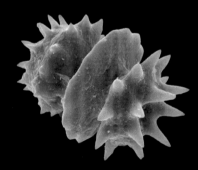

Various sponge spicules from Australia from the Late Eocene (40 million years ago). The elements are between 0.01 and 0.05 mm for the most part, except the two very elongated ones (top and bottom) which reach 0.2 mm.

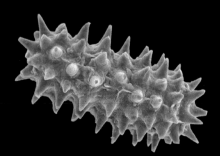

Illustrations of calcareous sponges and spicules by E. Haeckel, Plate 5, from his book *Kunstformen der Natur* (*Art Forms in Nature*), 1899–1904.

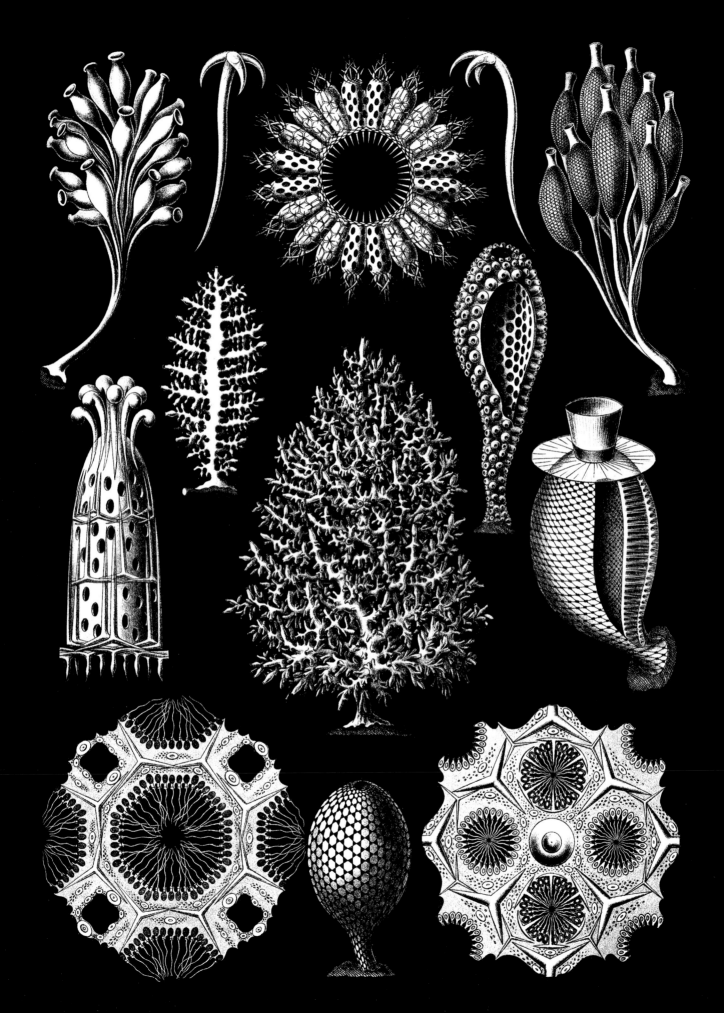

*An active mind tries to understand
and, inevitably, it doubts.*

Jean-Louis Fournier,
La servante du Seigneur
(The servant of the Lord)

Larval
Crumbs

Some marine animals have spiny skin, hence their name echinoderms, which include sea urchins. These spines of varying thinness and length are often inserted in fused plates, as in sea urchins, or articulated plates, as in sea stars, sea lilies and so on. In the absence of plates, the spines are supported only by an organic, collagenous substance, as in holothurians.

During their development, some echinoderms do not complete their morphological transformation until after they become sexually mature and can reproduce. Consequently, some larvae can reproduce. Occasionally, even females remain in the larval stage and are fertile, while males must finish developing before becoming fertile. This is the case for "glowworms": the female remains a larva all her life (hence the name "worm" even while a beetle), while the male must metamorphose into a winged insect before he can reproduce.

In echinoderms, this phenomenon is well known because holothurians, also called sea cucumbers, are larvae that have become fertile. They have soft, oblong bodies, but only small millimetric elements called sclerites are preserved during fossilization.

Illustrations of Thuroidea (holothurians) by E. Haeckel, Plate 50, from *Kunstformen der Natur* (*Art Forms in Nature*), 1904, showing various ossicles, or sclerites.

Conodonts'
Conical Teeth

In rocks there are sometimes small conical "teeth" whose colors range from opalescent white to dark brown, as well as all kinds of other colors—ivory, honey, amber and so on—depending on the temperature to which the rock was subjected (metamorphism). These colors are used to identify the past temperatures of the rock. Conodont elements, generally between 0.2 and 1 millimeter in length, do not remain the same over time. Therefore, they have been used to date rocks. However, it was not until nearly the end of the 20th century that we knew to which organism these elements should be attributed. The elements were named, like the rest of living things, with genera and species names, but they were only morphological species, which do not have the same meaning as biological species. When the organism with these elements was discovered around 1985, it became clear that only one organism possessed varied elements, and thus there were several morphospecies.

The "conodont animal" was discovered in 1983 by Euan Clarkson in Edinburgh in a drawer where it had been living for decades. It was a vermiform organism of a few centimeters with huge eyes. The elements, which had been known for a long time, corresponded to the mineralized parts (made of calcium phosphate) of the masticatory apparatus. The animal form of these conodonts is known to have existed since the Cambrian Period (about 500 million years ago), but it disappeared at the boundary between the Triassic and the Jurassic Periods, 200 million years ago.

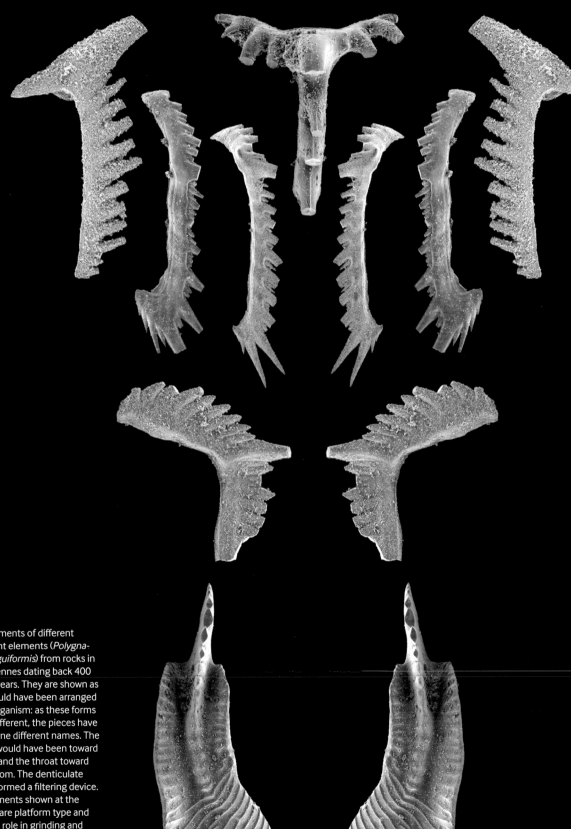

Arrangements of different conodont elements (*Polygnathus linguiformis*) from rocks in the Ardennes dating back 400 million years. They are shown as they would have been arranged in the organism: as these forms are all different, the pieces have long borne different names. The mouth would have been toward the top and the throat toward the bottom. The denticulate blades formed a filtering device. The elements shown at the bottom are platform type and played a role in grinding and ingesting food particles. The largest element is approx. 2 mm long.

144

Bottom:
These teeth form two prehensile sets on each side of the mouth. Chaetognaths are small, very ferocious predators (between 2 mm and 12 cm) that are extremely common in seas and oceans. Whole chaetognaths are found in Chengjiang (China). The strong resemblance between the elements in both rows suggests that some protoconodonts are in fact primitive chaetognaths.

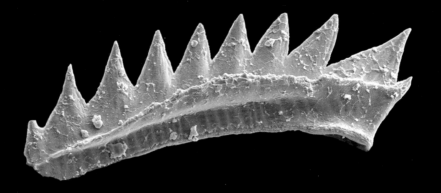

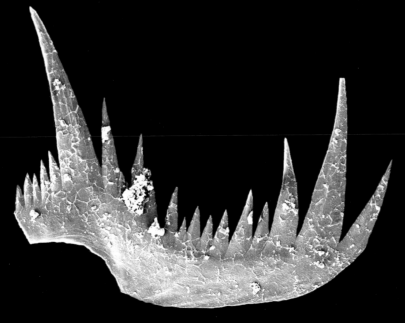

Various conodonts from
Turkey dating back
approx. 220 million years
(Triassic, Carnian).

The most beautiful thing we can experience is the mysterious.
It is the source of all true art and science.
He to whom this emotion is a stranger,
who can no longer pause to wonder and stand rapt in awe,
is as good as dead: his eyes are closed.

Albert Einstein

Sack-Shaped Animals?
Chitinozoans

Among the microremains found in rocks, there is a group of undetermined biological affinity made of organic material, more specifically of chitin, hence their name chitinozoans (from the Greek *khiton*, meaning "tunic," and *zōon*, meaning "animal"). Their size is between 0.05 and 1 millimeter. They are known to have existed since the Early Ordovician Period (about 485 million years ago), and in 2013, three small vesicles shaped like amphorae were found from the Middle Cambrian Period (510 million years ago). These microfossils emerged from 600 kilograms (approx. 1,323 lbs) of rocks when they were dissolved with acid—only three 0.6-millimeter fossil particles for 600 kilograms of rocks! Because their form varied over time, these microfossils are precious tools for dating rocks. They have also become tools for measuring the paleotemperature of seawater.

Chitinozoans occur as tiny vesicles, isolated or grouped in chains or clusters. They are shaped like urns, tubes or bottles. The tops of their tests are closed by a complex cap (operculum). It is assumed that they were produced in groups and that the vesicles were interconnected. They have been attributed to various organisms or to the many animals that have not yet been identified: worms (annelids) and gastropods have been good candidates for a while.

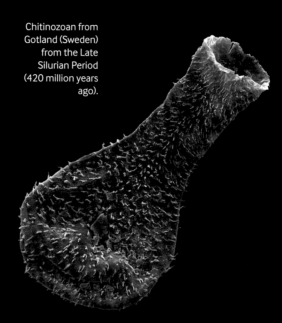

Chitinozoan from Gotland (Sweden) from the Late Silurian Period (420 million years ago).

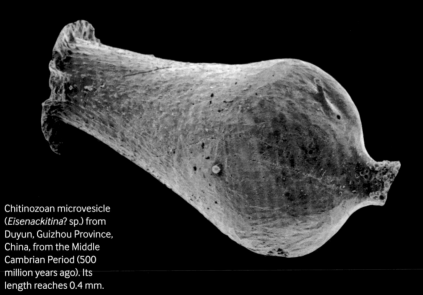

Chitinozoan microvesicle (*Eisenackitina*? sp.) from Duyun, Guizhou Province, China, from the Middle Cambrian Period (500 million years ago). Its length reaches 0.4 mm.

A flood of goldsmithery
Pierces the planktonic vault
And sifts the floor of stars
Tiaras and Eiffel towers.

Jean-Yves Reynaud

Radiolarians:
Polycystines and Acantharians

Radiolarians include two distinct planktonic groups: polycystines and acantharians. Consisting of a single cell with a nucleus, they are microscopic protists.

Acantharians build their skeletons out of strontium sulfate, a mineral that dissolves very quickly after the organism's death. Therefore, they are not really known in a fossilized state. On the other hand, most of the few thousand indexed species of polycystines build their skeletons, often in very elaborate forms, out of silica. They are known to have existed in rocks for 500 million years.

Some live in large colonies in a gel-like substance (cytoplasm), forming a translucent sphere or long cylinder that sometimes intrigues divers.

We find rocks made solely of agglomerated radiolarians, called radiolarites, at the tops of the highest mountains (such as the Alps and Himalayas). This attests to the fact that these mountains have marine origins (see "They Date the Opening," p. 204).

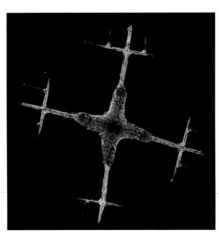

Juvenile acantharian radiolarian (*Lithoptera fenestrata*) living in the waters off the coast of Villefranche-sur-Mer, France (September 2014). Photo taken by differential interference contrast microscopy. The arms of the central part have a span of approximately 0.02 mm. In juveniles, the organic part of the body is smaller than in adults, but the skeleton is the same size, just less complete. It will be completed progressively as the individual grows.

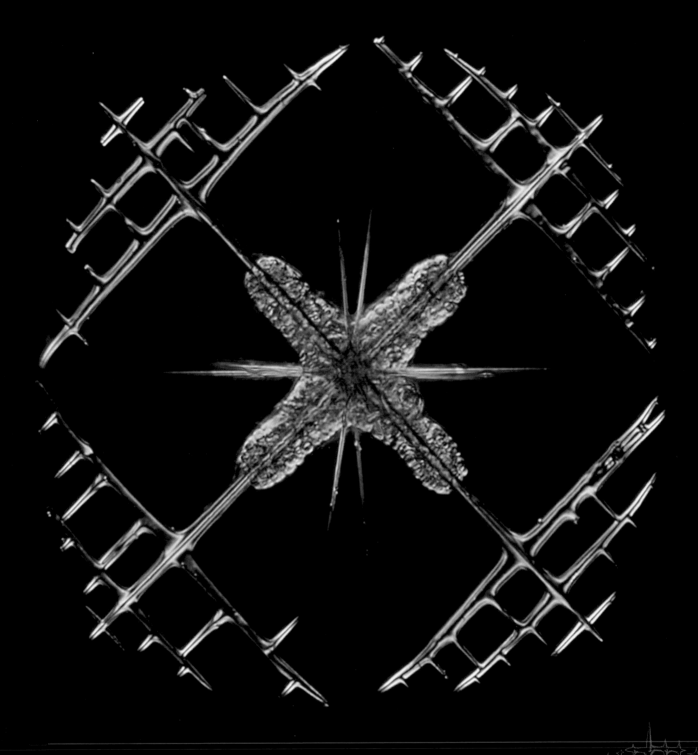

Acantharian radiolarian (*Lithoptera fenestrata*) living in the waters off the coast of Villefranche-sur-Mer, France (September 2014). Its skeleton is made of strontium sulfate, which grows with age. This half-millimeter unicellular organism is sustained by oceanic waters, where it lives for one to several months. The four yellow masses are symbiotic algae (*Phaeocystis* sp.) that live within the cytoplasm of the acantharian. The arms of the central part have a span of approximately 0.025 mm. Photo taken by differential interference contrast microscopy.

Which absentminded goldsmith
Which glassblower
Scattered his stuff
At the very bottom of the sea?

Jean-Yves Reynaud

Radiating
Radiolarians

Radiolarians are microscopic planktonic marine organisms that are found in all of the world's seas but not in fresh water. They live at all depths, even if the vast majority of them are found toward the surface.

A siliceous shell forms the single-celled (protozoan) organism's skeletal structure. The cytoplasm relies on the spines to produce filaments that move out from the central body. This strategy increases the radiolarian's chances of encountering inert particles or living organisms that it feeds on.

The skeleton is made of silica, which makes it resemble glass. Fossil forms retain only this mineralized part, so they look like lace made of crystal rock (the common name is quartz).

Known to have existed since the Cambrian Period (540 million years ago), radiolarians are among the oldest known groups of fossils on Earth. They are used for dating the rocks in which they are found.

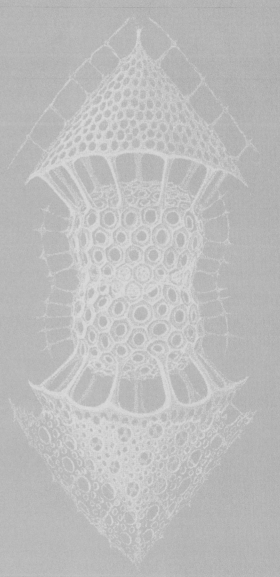

Siliceous plankton: illustrations of radiolarians published by E. Haeckel, Plate 91, group of spumellarians, from the book *Kunstformen der Natur* (*Art Forms in Nature*), 1899–1904.

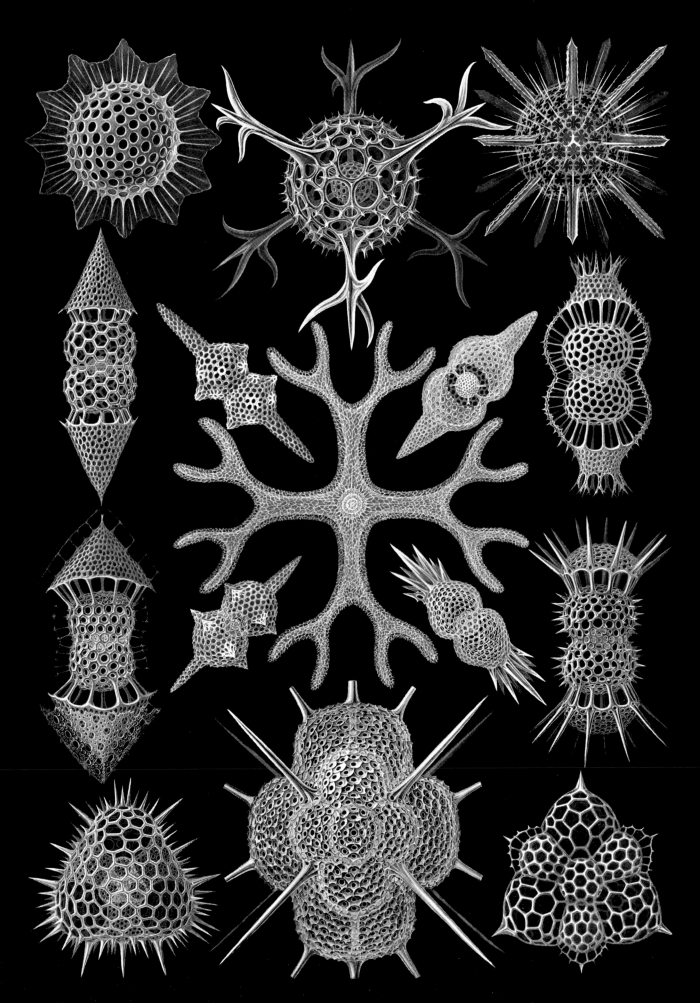

Oh sun!
You without whom things
Would not be what they are!

Edmond Rostand

Suns
That Enlighten the Geologist

Radiolarians are often in the form of a spiny mesh sphere, but they present a diversity of other geometrical forms, too. Their aesthetics undoubtedly place them among the most astonishing organisms.

Despite their spiky morphology, radiolarians are not ferocious hunters. However, if prey passes within their reach, they grab it with one of their many tentacles that can extend up to 10 to 20 or even 50 times the length of their test.

Radiolarians have long been considered hybrids, at once both animals and plants. Supporting the ambiguity of their nature, it must be stated that radiolarians generally need light to live, just like plants do. Indeed, they harbor unicellular algae even smaller than themselves called zooxanthellae, whose photosynthetic function they use in exchange for a certain amount of protection. These algae live in symbiosis with the radiolarian that hosts them and provide it with some of the metabolites necessary for its dietary needs, which allows them to fast for several days. For this reason, radiolarians are concentrated in the water close to the surface that is exposed to light.

Siliceous plankton: illustrations of radiolarians of the order Stephoidea published by E. Haeckel, Plate 71, from the book *Kunstformen der Natur* (*Art Forms in Nature*), 1899–1904. The small yellow spheres are photosynthetic algae.

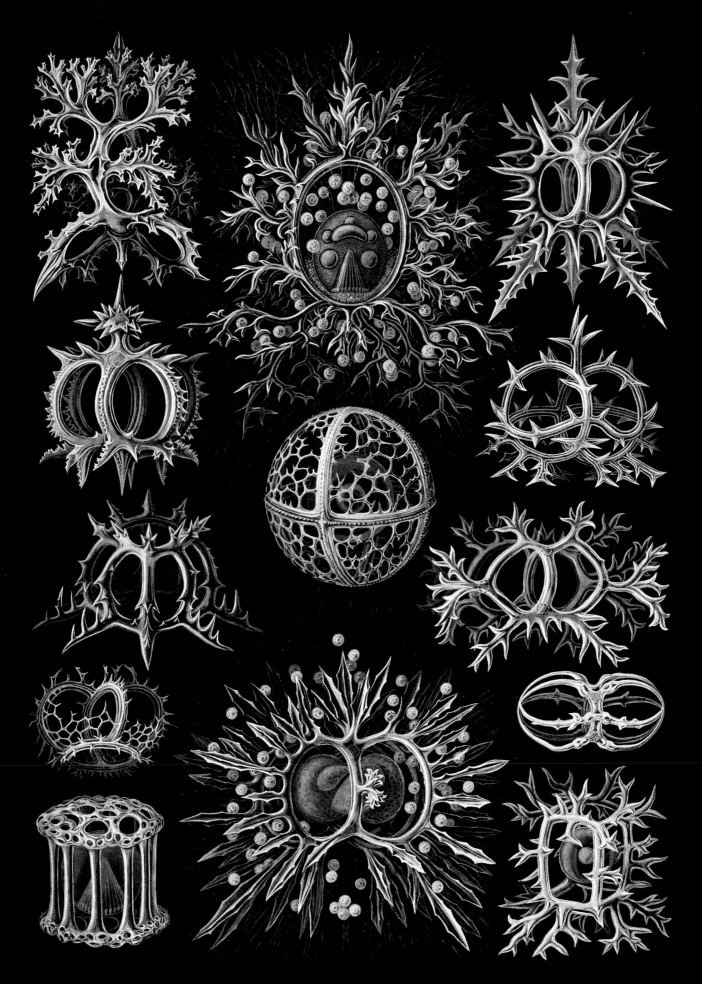

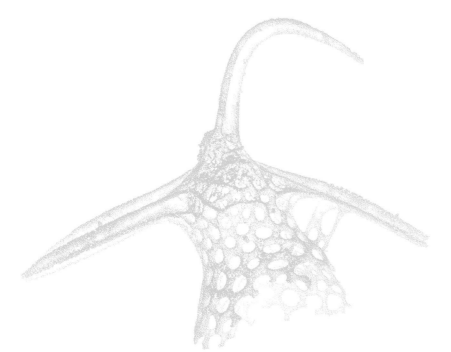

*Small is beautiful.
And besides there are so many . . .*

Jean-Yves Reynaud

Small, Abundant,
Varied

Through the diversity of their forms and colors, the micro-organisms swarming in the ocean always offer us new topics of interest and wonder. Among them, radiolarians are certainly the most elegant. Their thin latticed shells combine in nearly endless ways to form spheres, cylinders, cones, disks, rings, spirals and so on. And yet, as elegant as they are, radiolarians are among the marine protists about which biologists know the least. This lack of knowledge is simultaneously a result of their small size (from 0.005 to 0.3 mm on average), the relative complexity of their organization and, above all, the difficulty of cultivating and maintaining them.

For example, we have only known their eating habits since the end of the 20th century, as a result of observing them in the aquarium using high-speed cameras just after they were caught.

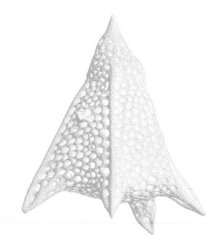

Various radiolarians of different ages showing the variety of their forms.

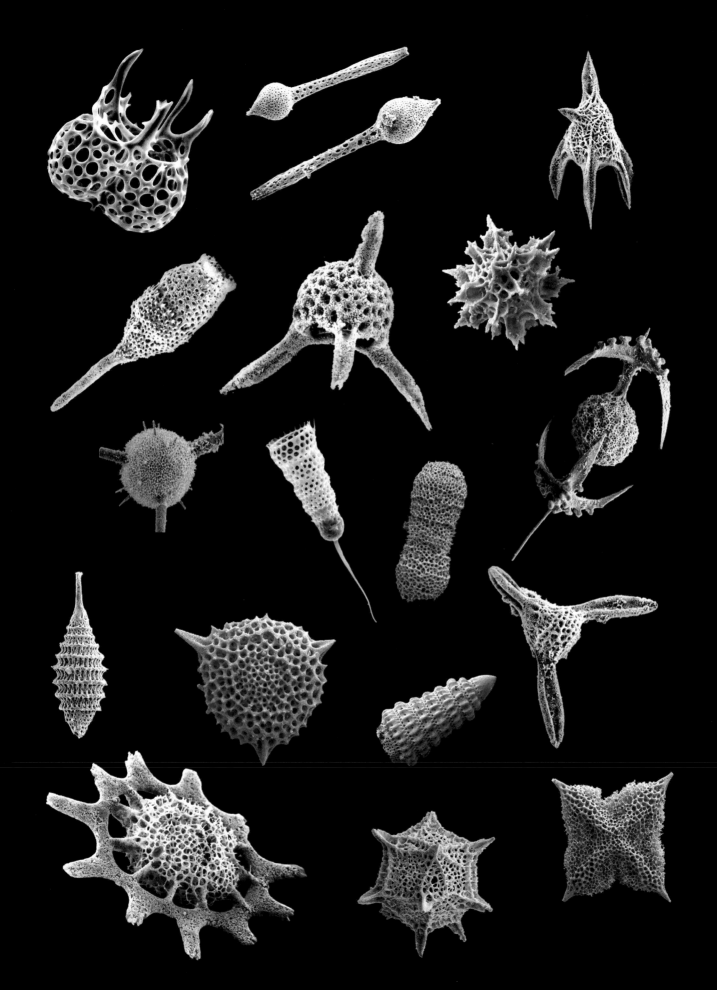

Behind their open windows the radiolarians
Keep the memory of solar air currents
But their small boat-shaped remains
Lack the zooxanthellae's lost sails.

Jean-Yves Reynaud

Floating in the Water
Transports You . . .

Radiolarians do not swim. They do not move horizontally. They let themselves be carried by the currents. They only know how to move vertically by means of their carbon dioxide–filled vesicles; to descend, they expel the contents. Radiolarians live on the surface to get light, but when it is too hot or when a storm makes the water too choppy, they dive. A little bit of sun, spending their time in the water fishing . . . These organisms seem to prefer a calm life—who could blame them? Many radiolarians live in colonies. Attached to one another by a kind of gelatin, they form colonies, the largest of which can reach 5 meters (approx. 16.4 ft). These voracious colonies feed on large prey (crustacean larvae, jellyfish, etc.).

Pelagic ooze, washed and attacked with acid, reveals small siliceous skeletons of radiolarians, sponge spicules and diatoms (Western and Central Pacific Ocean). These organisms will then be observed under an electron microscope or be beautifully drawn (plate by E. Haeckel from *Die Radiolarien: Eine Monographie* [Radiolaria: A Monograph], Berlin: Altas, 1862).

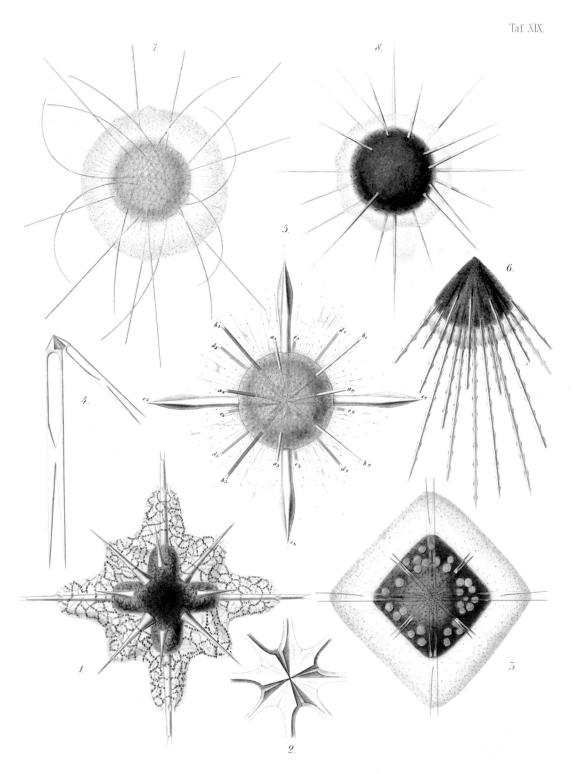

1-5. Acanthostaurus. 1. 2. A. purpurascens, Hkl. 3. 4. A. Forceps, Hkl. 5. A. hastatus, Hkl.
6. Litholophus Rhipidium, Hkl. 7. 8. Acanthochiasma. 7. A. Krohni, Hkl. 8. A. fusiforme, Hkl

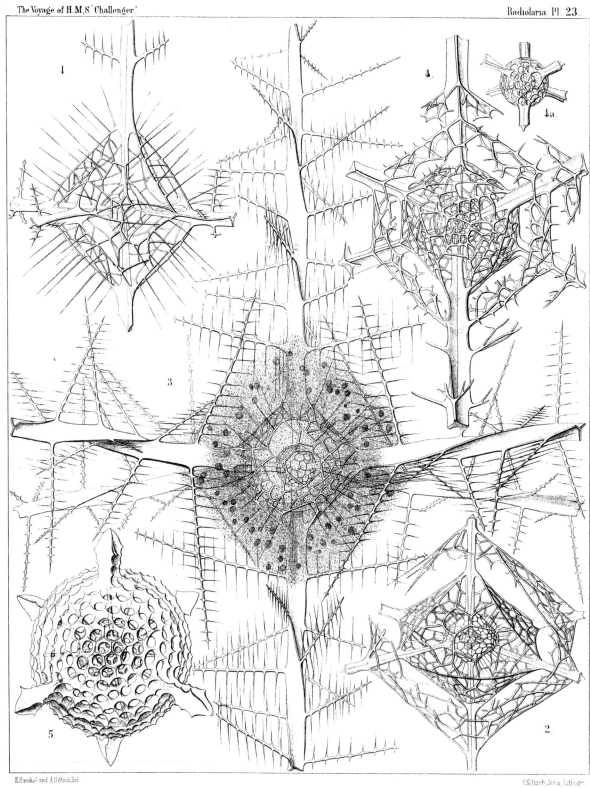

E.Haeckel and A.Giltsch,Del.

C.Giltsch, Jena, Lith.gr.

1.2. HEXADENDRUM , 3. HEXAPYTIS, 4. HEXACARYUM,
5. HEXACONTIUM.

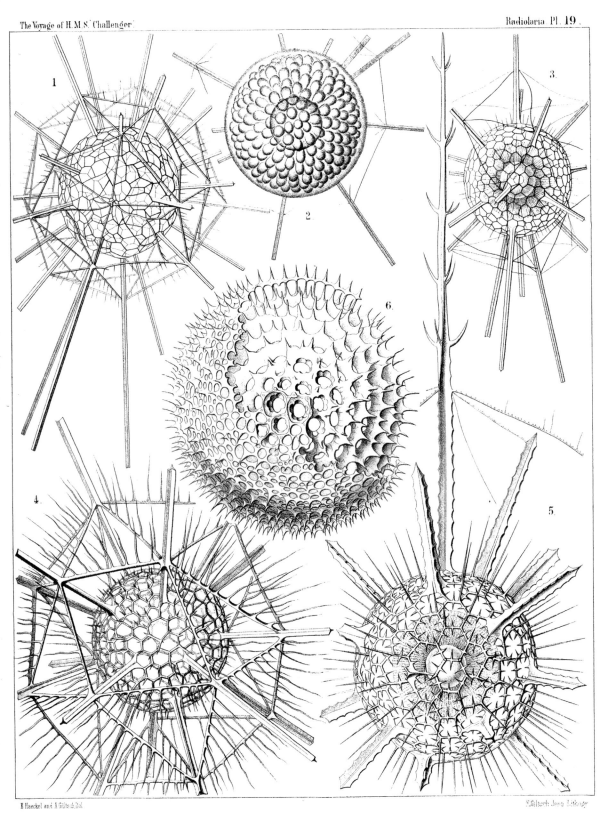

1-5. DIPLOSPHAERA, 6. SETHOSPHAERA.

The Diversity of Microfossils

The man who does not go out and visit
in all its vastness
the Earth full of its multitude of wonders
might as well be a frog in a well.

Indian Proverb,
excerpt from *Panchatantra*

Radiolarian:
The Perfect Body

On a daily basis, nature shows us many wonderful things, whether we look at them from our windows or from the top of a mountain. In a microscopic world—in other words, a world invisible to humans—beauty also has its place.

Could it be that a life in the seas trapped in the anonymity of invisibility would be compensated by the satisfaction of living in a body with a dreamy shape? In any case, this is the destiny of many microorganisms, including radiolarians.

These forms offer the eyes plentiful and extremely varied shapes and structures. A sort of boundless inventiveness. But as they are invisible, who is this for? For what purpose?

Maybe, quite simply, for living.

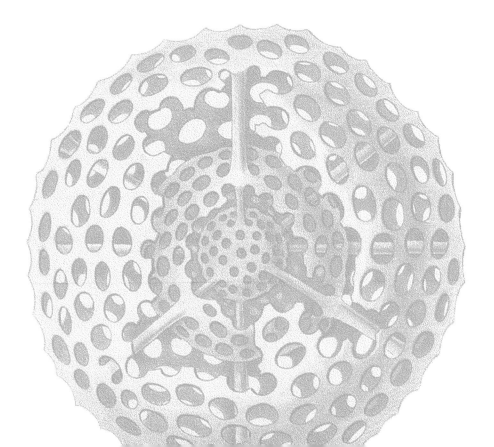

Illustrations of radiolarians (siliceous plankton) by E. Haeckel, published in 1862 (*Die Radiolarien: Eine Monographie* [Radiolaria: A Monograph]).

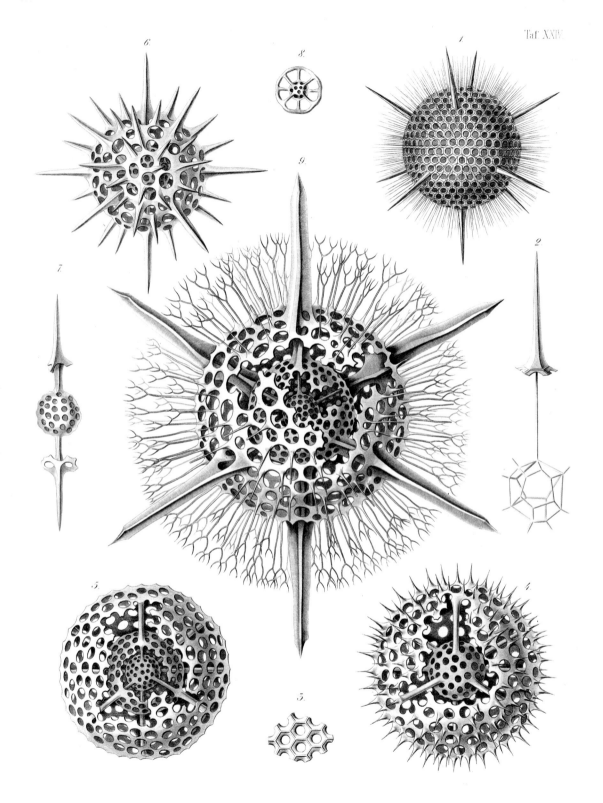

1-4. Hahomma. 1-3. H. Echinaster, Hkl. 4. H. Castanea, Hkl. 5-9. Actinomma.

5. A. inerme, Hkl. 6-8. A. Trinacrium, Hkl. 9. A. drymodes, Hkl.

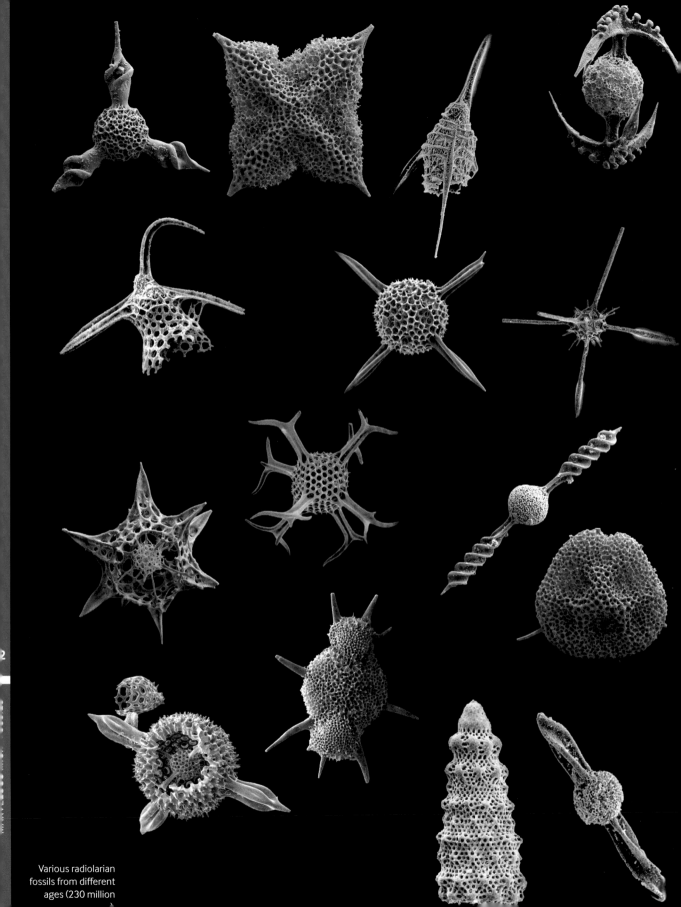

Various radiolarian
fossils from different
ages (230 million

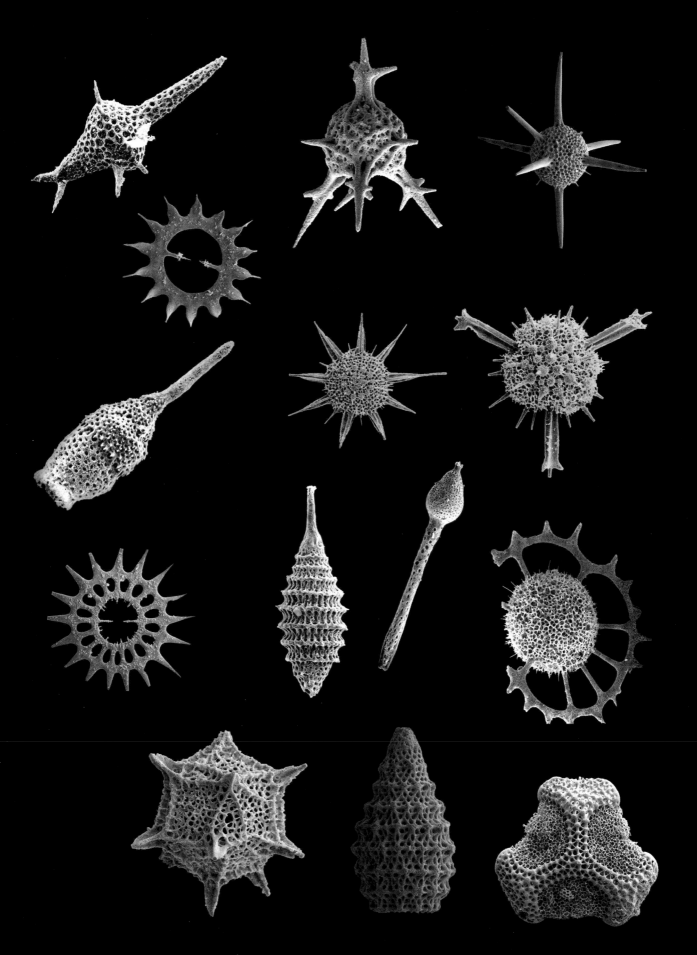

If you love a flower that lives on a star,
It is sweet to look at the sky at night.

Antoine de Saint-Exupéry,
Le Petit Prince
(*The Little Prince*)

Palynomorphs:
The Small Become Large

Some elements that micropaleontologists study are microscopic parts (between 0.002 and 0.2 mm) of macroscopic organisms, and sometimes they are very large organisms, like trees. Yet these giants use microscopic elements to reproduce, and when the season of love arrives for plants, their vital urges give insects and the wind billions of grains to scatter and reach a lucky ladylove.

A grain of pollen (from the Greek *palē*, meaning "flour," "dust") has a wall made of sporopollenin, which is probably the most inert and resistant material in the living world. This material does not age. Its resistance explains how it is preserved in sediments that have undergone many changes. Many sediments retain pollen grains without altering them for hundreds of millions of years. This distinguishing quality allows us to reconstruct forests from the Carboniferous Period and to date the rocks in which the grains are found. Spores and pollen grains are loquacious witnesses of life on Earth. These messengers are all the more important because there are very few traces of life on the continents due to a lack of effective recording mechanisms. Indeed, the continents are generally more affected by erosion than by sedimentation.

Palynomorphs from rocks from Archingeay in the Charentes-Maritime, France, dating from the Early Cretaceous Period (approx. 120 million years ago).

Top:
Pollen grains from an angiosperm. Size: approx. 0.035 mm. Image taken with a scanning electron microscope.

Bottom:
Trilete spores (Appendiscisporites). Size: approx. 0.04 mm. Image taken with a scanning electron microscope and colorized.

> *The known is finite, the unknown infinite; intellectually we stand on an islet in the midst of an illimitable ocean of inexplicability.*
>
> Thomas Henry Huxley

Nanoscale
Grape Seeds

The sizes of microplanktonic organisms vary by a factor of 1,000 to 10,000, meaning that these organisms come in all sorts of sizes. The smallest are elements from pico-plankton, which are unknown in a fossilized state. (They are generally less than 250 nm, or 0.00025 mm, though viruses are even smaller.)

These small organisms are called microscopic because they are smaller than a millimeter. When they are a thousand times smaller, they are called nanoscale. Fossils of this size are called nannofossils. They are a few micrometers or a few tens of micrometers in size.

These are, in fact, parts of organisms called coccolithophores, which have coccoliths (from the Greek *kokkos*, meaning "seed," *lithos*, meaning "stone," and *phoros*, meaning "which carries"). These organisms were described in 1836 by the German Christian G. Ehrenberg (1795–1876), one of the founders of micropaleontology. This is impressive, considering the limited means of observation at the time.

Coccolithophores are unicellular algae (protists) that live in the layer of water between 0 and 200 meters (approx. 656 ft) deep that is exposed to light, since they need light to synthesize chlorophyll. Their concentration in the water is variable but can be up to more than 100 million individuals per liter. They are among the great calcifying organisms of the living world and therefore are great regulators of the concentration of CO_2 in water and in the air, of the planet's carbon cycle and, for this reason, of the climate.

Coccoliths are known to have existed for nearly 220 million years (dating back to the Triassic, the first period of the Mesozoic Era).

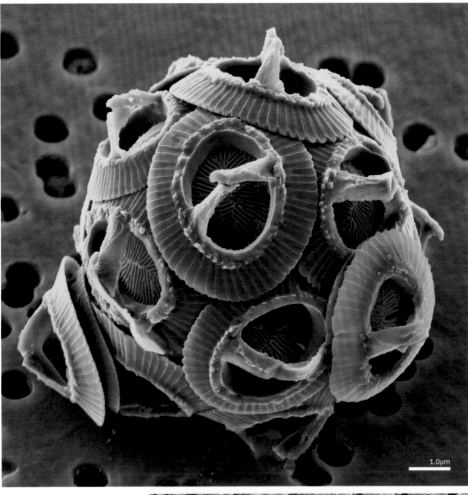

Coccolithophore (*Gephyrocapsa oceanica*) from the Mie Prefecture, Japan. The size of the coccosphere is 0.008 mm. Scanning electron microscopy, colorized photo.

Present-day coccolithophorid algae (*Emiliania huxleyi*) known in all oceans and extensively used to study changes in the climate. Upon an individual organism's death, the spheres usually break into small disks called coccoliths. The size of the coccosphere is 0.007 mm. Colorized photo taken with an electron microscope.

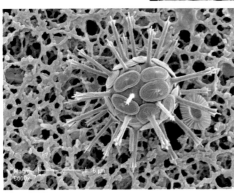

A coccosphere in its environment: a form with spines (*Rhabdosphaera clavigera*). In fact, two types of coccoliths are visible here. Some form an inner layer; these plates extend spines. Others form an outer layer that have no spines, only small buds. The diameter of the sphere is 0.007 mm.

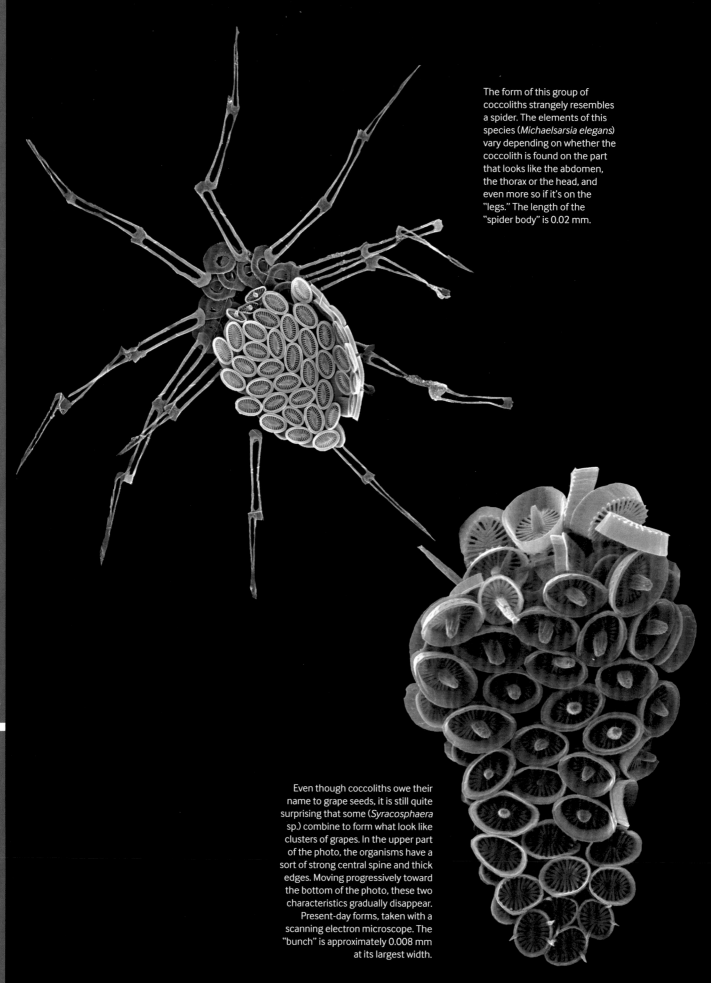

The form of this group of coccoliths strangely resembles a spider. The elements of this species (*Michaelsarsia elegans*) vary depending on whether the coccolith is found on the part that looks like the abdomen, the thorax or the head, and even more so if it's on the "legs." The length of the "spider body" is 0.02 mm.

Even though coccoliths owe their name to grape seeds, it is still quite surprising that some (*Syracosphaera* sp.) combine to form what look like clusters of grapes. In the upper part of the photo, the organisms have a sort of strong central spine and thick edges. Moving progressively toward the bottom of the photo, these two characteristics gradually disappear. Present-day forms, taken with a scanning electron microscope. The "bunch" is approximately 0.008 mm at its largest width.

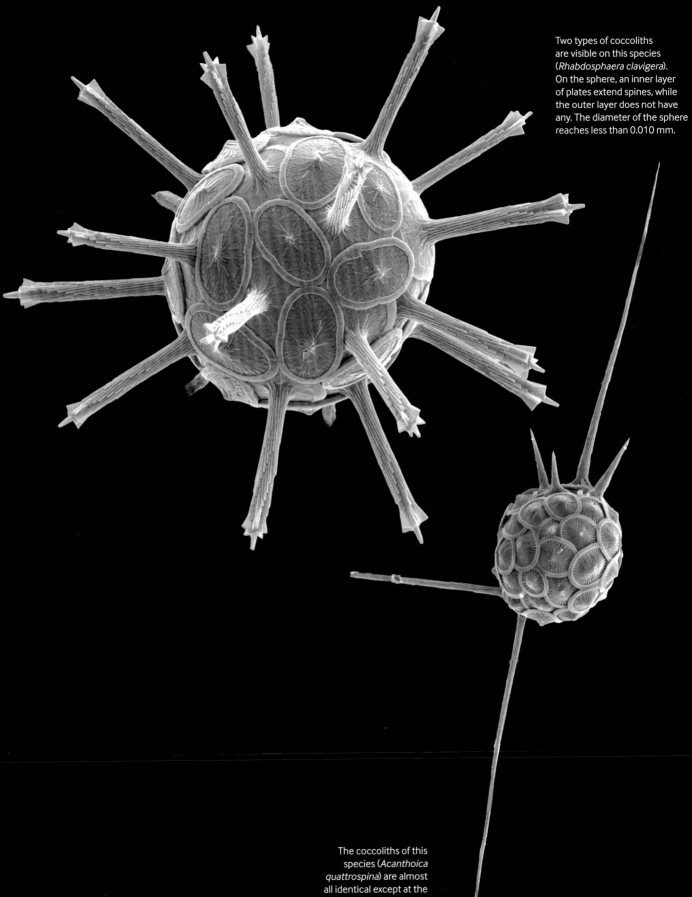

Two types of coccoliths are visible on this species (*Rhabdosphaera clavigera*). On the sphere, an inner layer of plates extend spines, while the outer layer does not have any. The diameter of the sphere reaches less than 0.010 mm.

The coccoliths of this species (*Acanthoica quattrospina*) are almost all identical except at the apexes, from which long spines soar. The width of the spherical part is around 0.006 mm.

Happy is he, like Ulysses, who had a good voyage,
Or like he who conquered the fleece,
And then returned, full of use and reason,
To live between his parents the rest of his years!

Joachim du Bellay

Ascidians

Ascidians are very special animals. Some are edible and are called "sea potatoes" (*patate de mer* in French) or "sea figs" (*figue de mer*) when they are purplish. They are the only animals whose heart alternates the direction in which it pumps, sending blood in both directions. They are also able to store vanadium, an element that is toxic for other organisms, accumulating more than 10 million times the concentration present in the surrounding seawater.

This group of marine animals is classified as tunicates or urochordates. They are considered to be an evolutionary group that is close to vertebrates. Their body, usually bladder shaped, is covered by a cellulosic sheath. Unusually for the animal kingdom, these animals are partially made of tunicin, a molecule similar to cellulose, which is characteristic of plants.

Ascidians have populated all of the world's oceans and are represented by more than 2,300 identified species. Sessile species colonize all environments, from depths of 400 meters (approx. 1,312 ft) to seaside rocks, and are even found on ropes in ports and on the bottoms of boats' hulls.

Only some small mineralized elements are found in the fossil record.

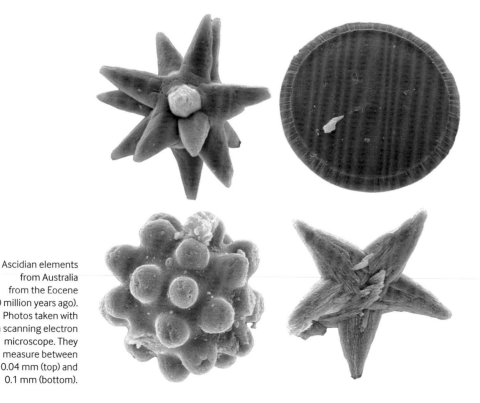

Ascidian elements from Australia from the Eocene (40 million years ago). Photos taken with a scanning electron microscope. They measure between 0.04 mm (top) and 0.1 mm (bottom).

Illustrations of ascidians by E. Haeckel, Plate 85, from the book *Kunstformen der Natur* (*Art Forms in Nature*), 1899–1904.

*Even dust, if piled up,
will become a mountain.*

Japanese Proverb

Phaeodarians

Little is known about phaeodarians. Their name comes from the dark granular pigments they contain (from the Greek *phaeos*, meaning "gray"). They are marine species that usually live quite deep in the water; a few species live at the surface. They are often solitary but sometimes gather together, showing the beginnings of organization.

The majority of species have internal skeletons. They consist of amorphous silica spicules. Sometimes the spicules are isolated. In some species, the skeleton is external, consisting of agglomerated foreign bodies (diatom shells, sponge spicules, etc.) on the periphery of the cytoplasm.

Skeletons of some phaeodarians consist of siliceous elements containing cytoplasm, connected to each other by a chitinous organic substance that disintegrates as soon as the organism dies. Therefore, it is problematic for them to be preserved in a fossilized state. Others consist of thin bars connected by reinforced knots or a very fine network much like that in diatoms. Finally, others have a gelatinous matrix with fine siliceous needles. Some species have bivalve shells.

Phaeodarians are known to have existed since the Cretaceous Period (70 million years ago in Japan; some rare specimens are known to have existed 95 million years ago in Russia).

Phaeodarians from rocks in Romania from the Middle Miocene (13 million years ago). Size: 0.2 mm. Photo taken with a scanning electron microscope.

Illustrations of phaeodarians by E. Haeckel, Plate 61, from the book *Kunstformen der Natur* (*Art Forms in Nature*), 1899–1904.

Bryozoans

Bryozoans (from *bryon*, meaning "moss," and *zōon*, meaning "animal," together meaning "moss animal") are colonial animals, most of which live in marine environments. These metazoans (multicellular) appeared in the Ordovician Period (450 million years ago).

Each organism makes a small protective covering out of chitin and most often lives within a sessile colony. In general, they are less than a millimeter in size, but encrusting colonies are visible to the naked eye. The colonies form a sort of whitish mossy carpet on the surface of rocks or any submerged object. Each organism has a crown of retractable ciliated tentacles. Most species produce a carbonate material that constitutes these protective coverings, and several species contribute to the construction of coral reefs. Some species are sometimes confused with corals. They contribute to the oceanic carbon sink along with corals.

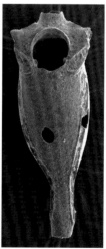

A bryozoan's protective covering. Its average width is approximately 0.14 mm. Photo taken with a scanning electron microscope.

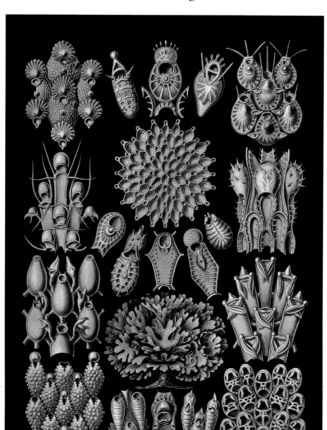

Illustrations of bryozoans by E. Haeckel, Plate 33, from the book *Kunstformen der Natur* (*Art Forms in Nature*), 1899–1904.

Illustrations of bryozoans by E. Haeckel, Plate 23, from the book *Kunstformen der Natur* (*Art Forms in Nature*), 1899–1904.

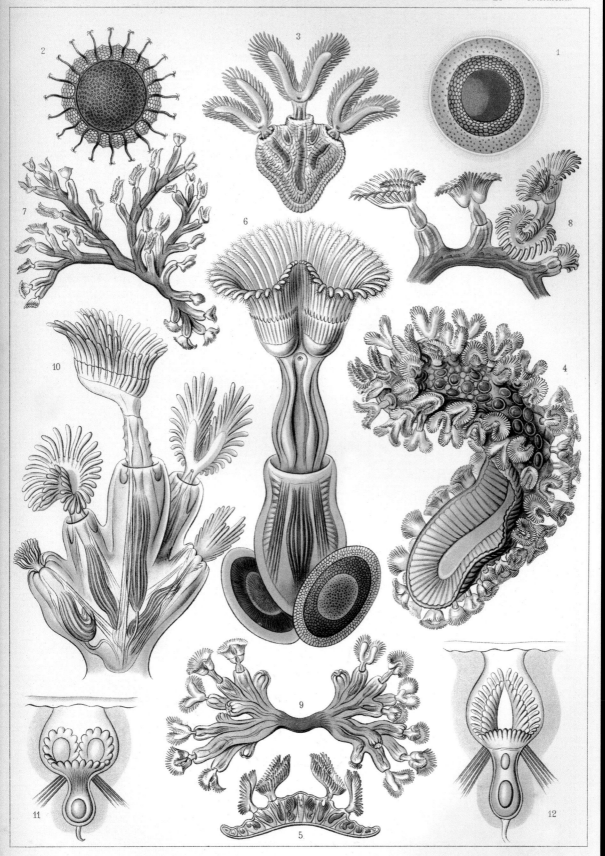

Bryozoa. — Moostiere.

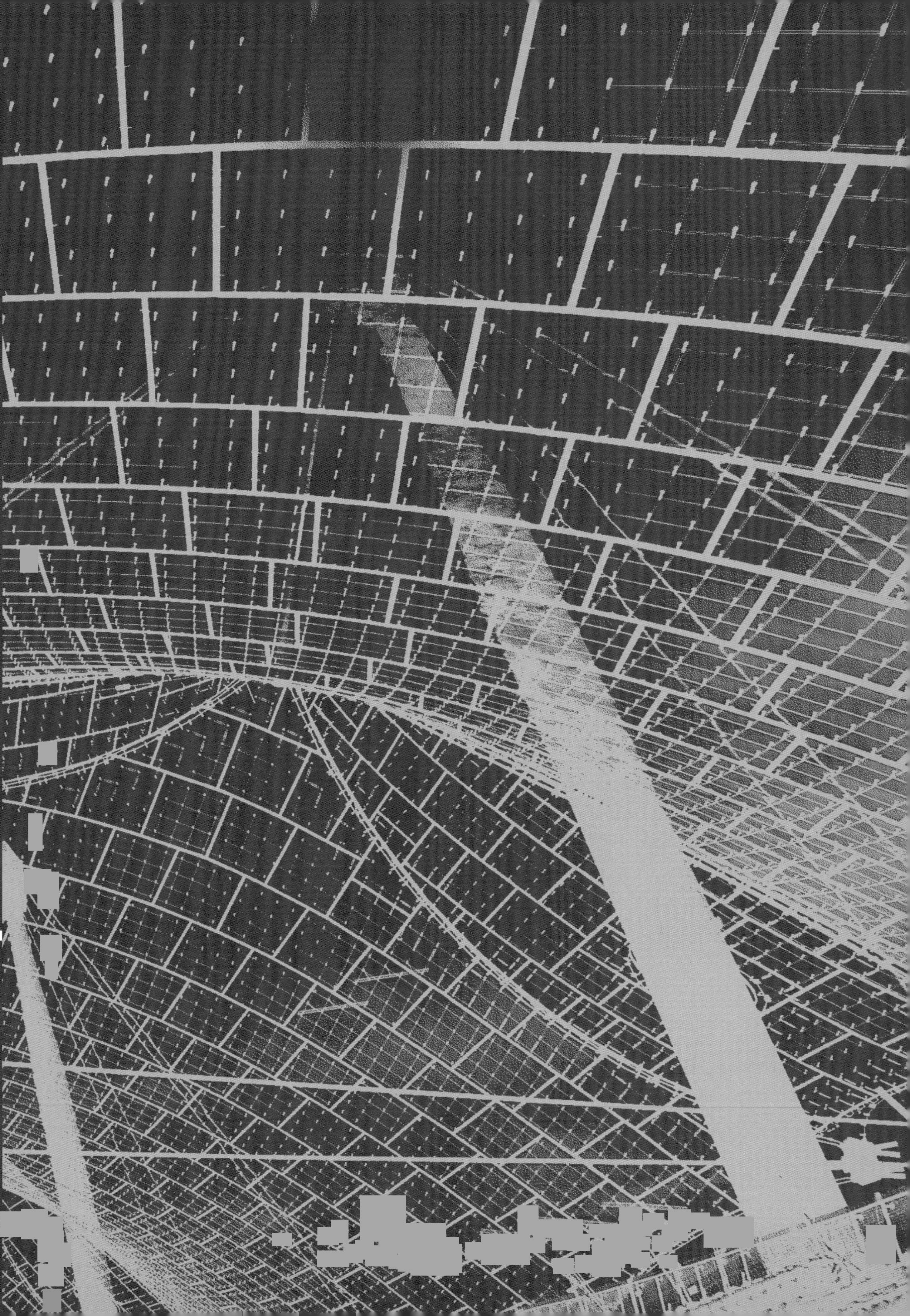

Architects, Creators and **Markers of Time**

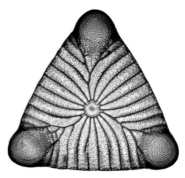

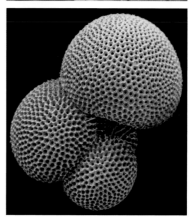

But if they cannot understand that parts so small that they are imperceptible to us can be as divided as the heavens, then there is no better cure than to make them look with glasses that magnify this delicate point into a prodigious mass; from which they will easily perceive, through the assistance of another even more artistically cut glass, that they might be enlarged to the point of equaling these heavens whose expanse they admire. And so, now that these objects seem to them so easily divisible, they remember that nature can do infinitely more than art.

Pascal,
Pensées (Réflexions sur la Géométrie en général)
(Thoughts [Reflections on Geometry in General])

The Optical Microscope
Opens Up New Horizons

The optical microscope was invented in 1690, but at that time it remained a rare object. Not easily accessed, it was as much an object of art as a tool of observation.

The world of microfossils was showcased in the 19th century thanks to the development of microscopes that facilitated the observation of the microscopic world. Until then, this world remained a mystery, as is always true of invisible things of which we are only mildly aware. People at that time tended to imagine a complex world consisting of a sort of chaos, surely not harmonious. Yet, the first surprise with the microscope was discovering that this was not so. This small world is remarkably organized, structured, precise and, moreover, strangely beautiful. The infinitely small intrinsically impose a change of scale. They also compel us to consider the fossilized skeletons of microscopic animals, for example the 190-million-year-old radiolarians, to be sources of inspiration as much for architects as for jewelers. Microscopes alone allowed us to discover this new world.

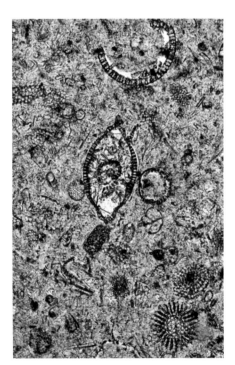

Very thin strip of rock (0.03 mm) that, held up to light, allows the microfossils within it to be seen; here are radiolarians that are approx. 40 million years old (Eocene). Sample taken during a deep sea drilling project in the Caribbean (Barbados). Their size is between 0.1 and 0.2 mm.

The surprising forms of microscopic organisms are glorified in very different works: motifs for fabric, paintings, sculptures, silver jewelry, belt buckles, rings and so on. I chose only to present the works of contemporary artist Michael Foster. They are all inspired by siliceous planktonic marine organisms (radiolarians or diatoms). See breezyhillturning.com

Nachet and Son microscope from the late 19th century.

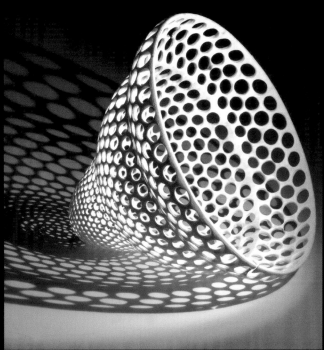

Reproduction of a radiolarian (*Dictyophimus hertwigii*); turned work, painted maple, inspired by E. Haeckel's illustrations, 3.5 in (height) × 4.75 in (diameter) (approx. 9 × 12 cm).

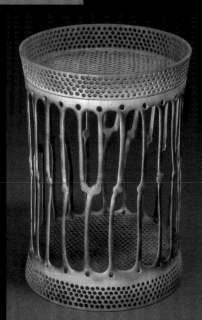

This work inspired by a diatom (*Triceratium* sp.) posed many technical problems for the artist (Michael Foster, dentist and woodturner). So many challenges, in fact, that he made it a point to confront them. Made of painted maple, 6.5 in × 2.25 in (approx. 17 × 6 cm).

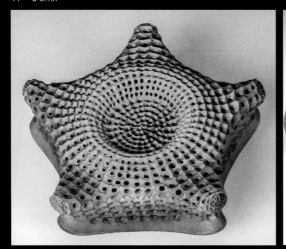

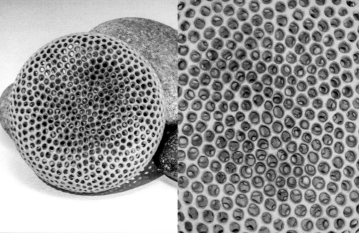

Piece inspired by a diatom (*Coscinodiscus* sp.). The blue evokes the sea where diatoms live and the green the chlorophyll they synthesize. Made of hornbeam, 5.5 in × 1.5 in (approx. 14 × 4 cm).

The problem posed by this piece based on a diatom was that while it appears that the structure is hexagonal, a sphere requires some pentagons to be inserted as the architect Buckminster Fuller did to build his domes (see "Very Modest Sizes," p. 248). Its diameter is 4 in (approx. 10 cm).

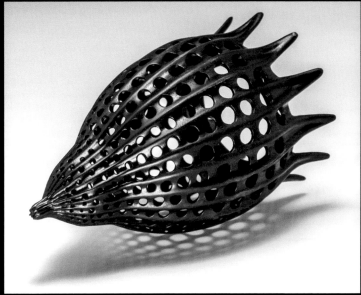

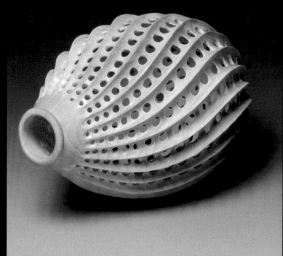

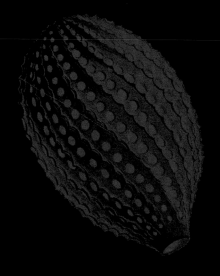

Top, from left to right:
Reproduction of a cyrtoidea
(radiolarian). Made of ebony,
3.75 × 6 in (approx. 10 × 15 cm).

Piece inspired by a radiolarian
(*Cyrtophormis spiralis*) made of
birch, 4 × 7 in (approx. 10 × 18 cm).

Bottom, from left to right:
Reproduction of a radiolarian
(*Astrocyclia solaster*); decimetric
piece made of ebony, 3 × 5 in
(approx. 8 × 13 cm).

The latest radiolarian creation by
this artist posed many problems
because of the delicately thin,
intensely perforated wall. Made
of painted maple, 12.5 × 5 in
(approx. 32 × 13 cm).

Who will probe this universe
And the powerful attraction that it deserves?
Come oh Newton of the human soul
And all the heavens will open!

Sully Prudhomme

Observing a Microscopic World
and Creating

The development of high-performance microscopes that were accessible to more people opened new horizons and allowed the imagination to run wild. Some curious people from diverse backgrounds—a few scientists, a few artists—were pleased to make such discoveries. Some were inspired by the elements they discovered, such as organisms or parts of organisms (see p. 20). From these small "bricks" they imagined and created microscopic mosaics. Each element was chosen, taken and positioned with precision and meticulousness attesting to great skill. Yet these pieces are only visible to a few "elect," or those who are insiders in some way.

How these observers approached the almost invisible evokes the Zen Buddhist monks who strive to create perfect lines in their garden by meticulously raking the gravel even when they know the design will be fleeting. Without fail, birds will come ruin this perfection as soon as the monks' backs are turned. It is precisely this brevity that makes their gesture beautiful: a long quest for perfection that will last a mere instant. They put all their soul, all their skill and their art into creating pieces that such a small number of people will be able to enjoy.

Top:
Microtableau of approximately 2 mm. This rosette is almost invisible to the naked eye. It consists of hundreds of skeletons of siliceous microalgae (diatoms). Each round or elongated element is the skeleton of an organism, measuring just a few tens of micrometers (between 0.02 and 0.2 mm).

Bottom:
Glass model of a radiolarian (*Aulosphaera elegantissima*) by Leopold and Rudolf Blaschka, 1885 (Museum of Zoology in Tübingen).

You see that whiteness that we call the Milky Way; what is it, would you say?
an infinity of small stars invisible to the eye because of their smallness,
and scattered so closely together that they seem to glow continuously.

Fontenelle

White Cliffs
of Manure

In our culture white is considered to be the symbol of purity and virginity. This is illustrated by the tradition of bridal gowns. Black, on the other hand, is the mark of the subterranean world, of the devil and of waste. For this reason no doubt, and perhaps also because it evokes the innocence of our childhoods and memories of the whiteness of chalk on the board at school, chalk itself is thought of positively. And yet this writing stick, before being cast, was made of the rock that gave it its name: chalk. This rock is made from material that is the result of processes that are far from being associated with purity and cleanliness. Indeed, this rock consists of the slow accumulation of small shells. Most of them were skeletal parts of "micro-microscopic" (nanoscale) algae that were eaten and whose mineral skeletons were discharged as fecal pellets.

The breathtaking cliffs in Étretat, France, are white, and yet it is essentially manure that gives us this lovely view!

A fecal pellet contains up to 40,000 components. These small feces sometimes constitute the bulk of a rock. Chalk is one such rock made of fecal pellets. There are several million individual pellets per cubic millimeter. Therefore, the very white chalk cliffs in Étretat, France, are, in fact, like a gigantic pile of manure!

Planktonic nannoalga (*Coccolithus pelagicus*). This sphere is the alga's skeleton (called a coccosphere); it consists of small disks, which are formed by plates arranged in a circle. The sizes of the disks range from 0.001 to 0.01 mm.

Planktonic nannoalga (*Discoaster surculus*). The size varies from 0.005 to 0.04 mm.

A sense of wonder guarantees happiness because life, if we know how to force its hand, never refuses man the opportunity to be amazed.

Ginette Quirion

So Small, Yet So Plentiful:
Plankton

The word *plankton* comes from the ancient Greek *planktos,* meaning "to wander, drift." It refers to all aquatic organisms (plant and animal) that drift with the currents. Generally microscopic or small in size, plankton do not truly swim (swimming organisms are called nekton). Plankton is the main source of food for baleen whales and filter-feeding shellfish, including mussels, cockles, oysters and so forth, which are sometimes poisoned by the toxins plankton produces.

Billions upon billions of plankters populate our oceans, seas, lakes, rivers and streams. Some drift between bodies of water, while others are attached to the bottom.

Even though the vast majority of plankton is invisible to the naked eye, its thousands of different species represent most of the biomass of marine organisms. In fact, plankton constitutes 98% of the mass of the sea's inhabitants. Crustaceans, fish and whales only represent 2% of living matter. Phytoplankton alone make up approximately 50% of the organic matter produced on planet Earth.

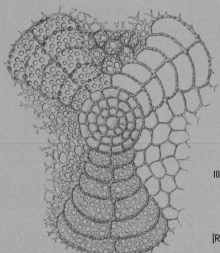

Illustrations of radiolarians (siliceous plankton) by E. Haeckel, published in 1862 (*Die Radiolarien: Eine Monographie* [Radiolaria: A Monograph]).

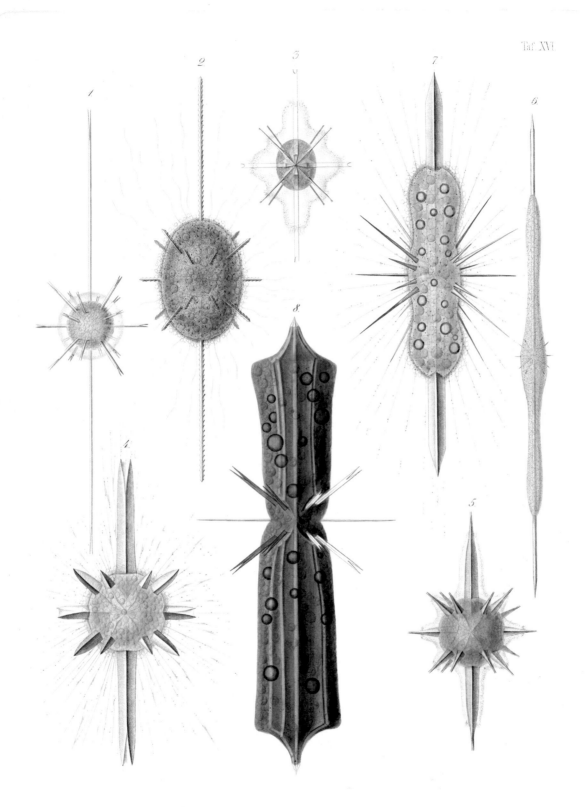

1–8. Amphilonche. 1. A. tenuis, Hkl. 2. A. denticulata, Hkl. 3. A. complanata, Hkl. 4. A. Messanensis, Hkl.
5. A. tetraptera, Hkl. 6. A. belonoides, Hkl. 7. A. heteracantha, Hkl. 8. A. anomala, Hkl.

E. Haeckel del.   Wagenschieber sc.

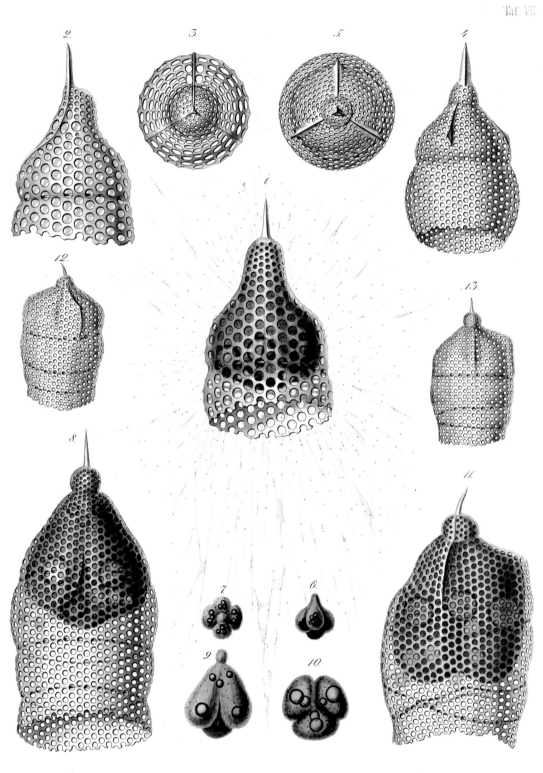

1–13. Eucyrtidium. 1–3. E. cranoides, Hkl. 4–7. E. carinatum, Hkl.
8–10. E. Galea, Hkl. 11–13. E. anomalum, Hkl.

E. Haeckel del.   Wagenschieber sc.

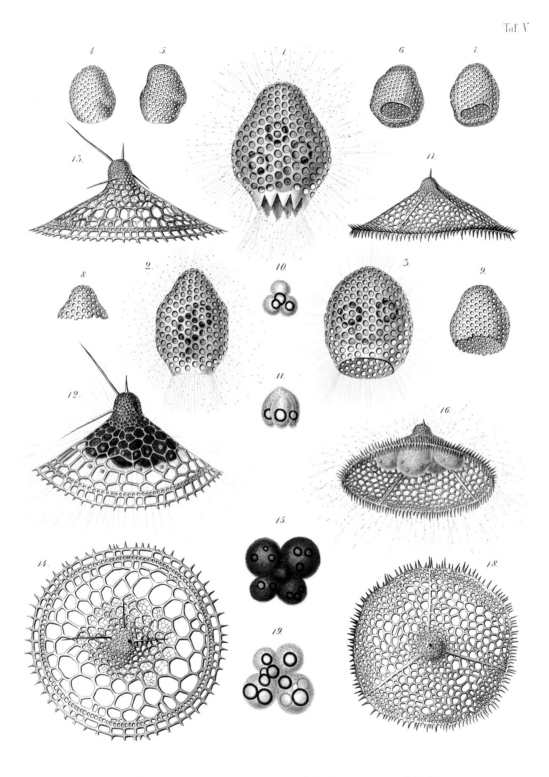

1. Carpocanium Diadema, Hkl. 2–11. Cyrtocalpis. 2. C. Amphora, Hkl. 5–11. C. obliqua, Hkl.
12–19. Eucecryphalus. 12–15. E. Gegenbauri, Hkl. 16–19. E. Schultzei, Hkl.

Microfossils
Everywhere

Every body of water, whether fresh, brackish, acidic, salt or frozen, contains a lot of both unicellular and multicellular organisms that float at the water's whim. We call this plankton. They exist in baffling numbers. For example, there are 10 to 100 million protists per liter of seawater. The aquatic ecosystems of rivers, lakes, glaciers, coasts and oceans each have their own dynamics. They change with the seasons, reorganizing themselves according to various currents, geographies, depths, available nutrients, climatic variations and so on. The dynamics of life cause this world to be in a state of perpetual change, affecting both the quantity and variety of individual organisms. These successive changes are recorded in the skeletons that microorganisms leave behind for us, which accumulate at the bottom of these bodies of water after their death. These mineralized remains go through time being piled into a sedimentary ooze that, after compaction, hardens and becomes rock.

Calcium and carbonate are abundant in seawater. Most organisms have skeletons (made of calcium carbonate), including many mollusks that are as large as the shellfish on our beaches or as small as the minuscule pteropods. The same is true for protists (foraminifers and coccolithophorid algae). Silicon is more rare, but it is nevertheless found in the skeletons of diatoms, heliozoans, silicoflagellates, radiolarians and so on.

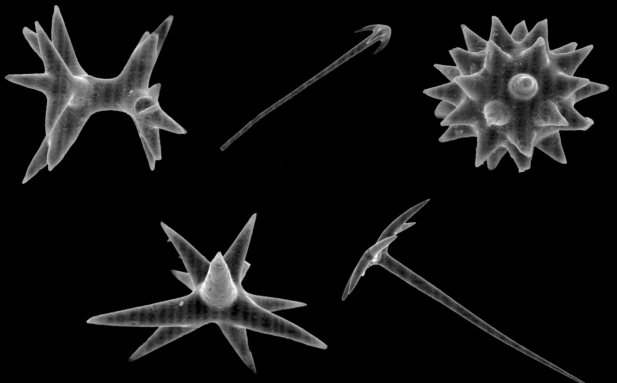

Sponge spicules from sediments in the Vienna Basin, Slovakia, dating from the Miocene (10 million years ago). Photos taken by scanning electron microscopy. The elements measure between 0.05 and 0.5 mm for the more tapered ones.

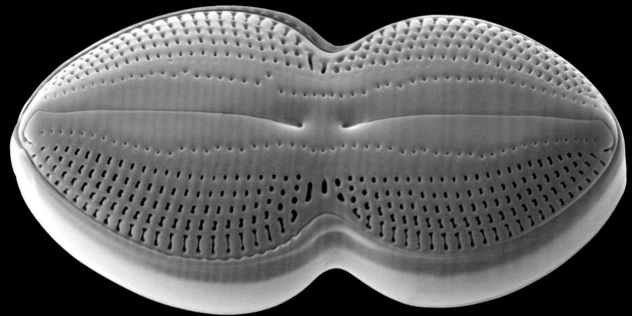

Small alga from present-day cold oceans: diatom. Note the delicacy of the perforations. Photo taken by scanning electron microscopy. The longest length is approximately 0.075 mm.

Bacteria:
Ubiquitous, Influential

Bacteria, archaea and viruses are present everywhere on the continents, in fresh water and in the oceans. On their own or in symbiosis, they occupy the water column and cover the surfaces of organisms' cells and tissues. Bacteria and archaea are also abundant in the bowels of organisms and in their excrement. Their sizes vary from a fraction of a micrometer (one millionth of a meter) for an individual bacterium to several millimeters for bacterial aggregates that form films or filaments.

Whether in oceanic "deserts" or blooms, bacteria are counted by the millions and billions in every liter of water! They are at the heart of life's symbiotic system.

Bacteria and archaea were the first living creatures to populate the Earth. For nearly two billion years, these were the only organisms that existed, deriving their energy from oxidizing metals and solar radiation. Bacteria and archaea gave birth to cells with nuclei and organelles called eukaryotes, the ancestors of protists, animals, plants and so forth.

Bacteria and archaea shaped the primeval planet, changing the composition of the water and atmosphere just by pumping carbon dioxide and releasing oxygen.

Recent stromatolites from Lake Alchichica (Chile). The rocky elements are stromatolitic structures.

Sectional image of a present-day microbialite showing cyanobacteria. The cyanobacterial sheaths stand out in green. Photo taken by confocal laser microscopy. Sample from ponds in Narbonne, France.

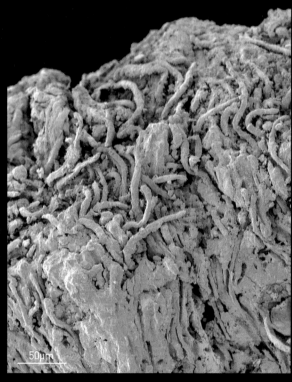

Cerebroid microbialite. In this recent formation, the encrusted filaments are subparallel and deep (bottom right) and also less organized and on the surface (top left). Photo taken with an electron microscope. Pauli Mesa Longa, Romania.

50μm

50μm

Astonishing voyagers! What remarkable stories
We read in your eyes as deep as the seas!
Show us the chests of your rich memories,
Those marvelous jewels, made of stars and ether.

Baudelaire,
"Le voyage"

So Small
but So Important

The planktonic ecosystem, a vast community of living things drifting with the currents, is a true ode to life. Life appeared more than 3.5 billion years ago, and ever since, life has not ceased to reproduce itself. Because each offspring is not exactly identical to its parent, some life-forms have adapted better, and others less so. Evolution is the cumulative process of these successes and failures.

The diversity and abundance of plankton vary with the currents, geography and the seasons. Some explosions of life cause blooms. They are likely to devastate aquaculture, to initiate the formation of clouds or to color and illuminate the sea.

Plankton, the primary food source for all marine organisms, is the initial, and therefore essential, link. The marine food pyramid shows that 1,000 grams of phytoplankton feed 100 grams of zooplankton, which then feed 10 grams of fish fry and crustacean larvae (juveniles), which feed 1 gram of "forage" fish (small fish), and this gram feeds 0.1 grams of tuna. In other words, it takes 10 metric tons of phytoplankton to produce 1 kilogram of tuna!

Illustrations of radiolarians (siliceous plankton) by E. Haeckel, published in 1862 (*Die Radiolarien: Eine Monographie* [Radiolaria: A Monograph]).

1–5. Dictyoceras Virchowii, Hkl. 6–10. Dictyopodium trilobum, Hkl.

Plankton
Manages the Climate

Life is influenced by the conditions that exist on the surface of the Earth, and conversely, life partially manages how the planet functions. For example, the release of oxygen produced by the first microscopic organisms changed the composition of the ocean and the atmosphere. Life participates in the self-regulation of environmental conditions.

Dark regions, such as mountains in the summer, forests or the ocean, tend to absorb the sun's energy. Conversely, bright regions, such as deserts or polar ice caps, reflect energy. Reflection is also influenced by cloud cover, which is itself dependent on the emission of gas by certain groups of phytoplankton, such as coccolithophores (the algae that formed chalk) or *Phaeocystis.*

For clouds to form, an aerosol particle must be present to collect water into a droplet. Dimethyl sulfide is one such substance that does this. Planktonic algae produce this gas, thereby forming clouds. They have the Earth's thermostat "in their hands"! When the sun shines, phytoplankton grow and release this gas in quantity, producing clouds. These clouds reduce insolation and consequently the temperature, which also slows the growth of phytoplankton and the amount of gas emitted. Then, there are fewer clouds and the temperature rises once again. The cycle continues in a dynamic, self-regulating and stable manner.

During this storm in Iceland, a yellowish foam accumulated in the basalt rocks. It consisted of microalgae, called *Phaeocystis,* which play a big role in the formation of clouds.

Collodarians, radiolarians (*Thalassolampe margarodes*). Around the central capsule (bright ocher color) large vesicles regulate buoyancy and constitute lipid nutrient reserves. The many yellow particles are photosynthetic microalgae (zooxanthellae) that live in symbiosis. Plankton collected off the coast of Villefranche-sur-Mer, France.

Skeletons
for Filtering Wine

When organisms die, they are deposited in a thin film at the bottom of the water. Their organic matter is destroyed, but their skeletons settle to the bottom and can accumulate in thick layers that are then exploited by industry. Rock made only of this accumulated sediment is called diatomite. Made of perforated particles, the rock is very finely porous and therefore used for filtration. The neutrality of silica makes it an excellent filter, and so it is widely used in agriculture. For example, it is now used to filter wine whereas the old "fining" technique utilized egg whites.

The rock deposits formed by these skeletons are generally brittle rocks that are also widely used in agriculture to lighten the soil, in electrical insulators, as a filler in paints, for anti-abrasive coatings or, on the contrary, as abrasives such as in some toothpastes.

Its porosity also makes it a natural container for petroleum (in California). Another substance that makes use of this property is dynamite. A stick of dynamite is a cylinder of diatomite soaked in trinitroglycerin (TNG), thus allowing it to be transported without the risks illustrated in the film *The Wages of Fear*.

This rock's high porosity gives it very low density, which is useful in some types of construction (see "Very Modest Sizes," p. 248).

Photosynthetic siliceous alga: diatom (*Triceratium polycystinorum*) from sediments in Russia dating back approx. 55 million years (Paleocene-Eocene). Photo taken by optical microscopy. Size: approx. 0.15 mm.

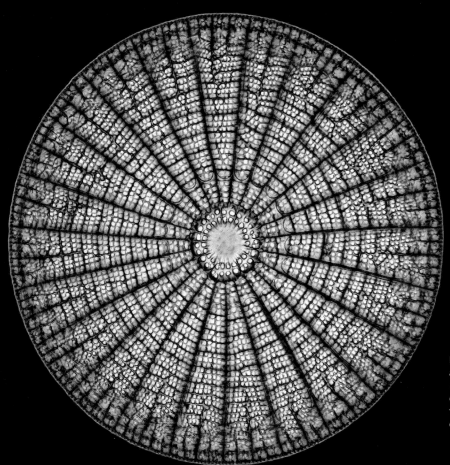

Photosynthetic siliceous alga: diatom (*Arachnoidiscus* sp.) living in cold waters off the coast of Peru. Photo taken with a scanning electron microscope.

Isthmia, Glorioso Islands. Details of one specimen. Size: 0.005 mm.

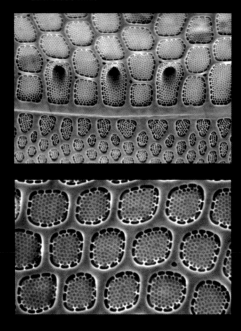

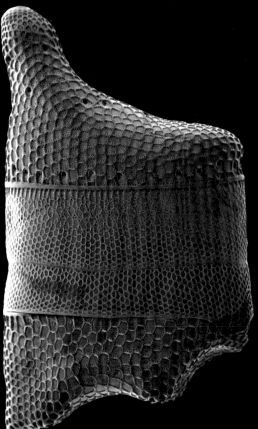

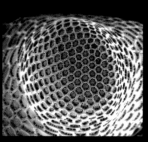

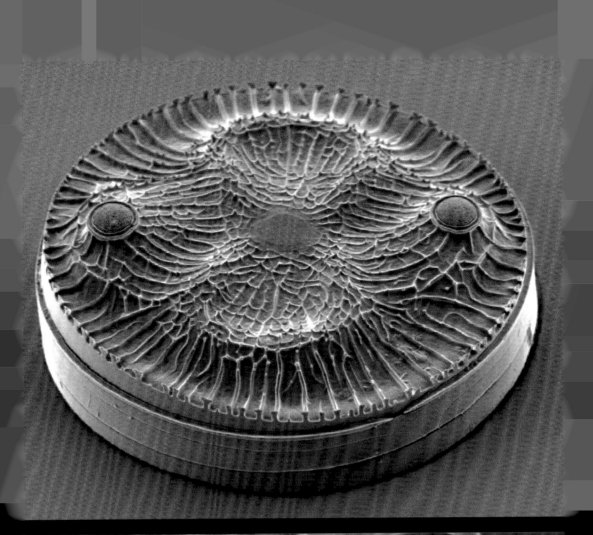

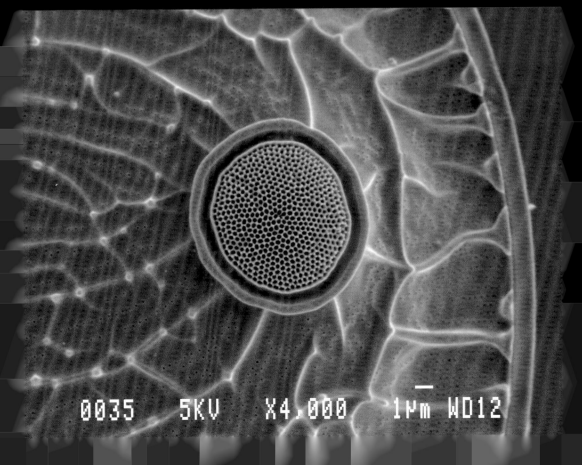

0035   5KV   X4,000   1μm WD12

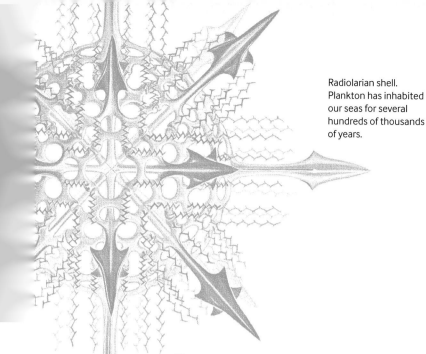

Radiolarian shell. Plankton has inhabited our seas for several hundreds of thousands of years.

Fossil
Timekeepers

Life is affected by evolution. How it functions, how it is organized and the forms it takes have therefore all changed over time. Organisms have left us firsthand memories of their lives, such as their skeletons, or incidental ones, such as their traces and things they built (e.g., reefs and stromatolites). First, this is enough to establish the chronology of these elements, to establish a scale, and then to use these elements for dating other things. We use the same approach as the child who dates a time in human history by the clothing: the toga of the Romans, the tricorn hat of the 18th century, the top hat of the 19th century and jeans with holes for the 21st century.

Some organisms had a short life span, during which they were found everywhere. This makes them good indicators of time. Among these are many microfossils (foraminifers, radiolarians, etc.). In fact, unicellular organisms generally have a fairly short life cycle, so during their successive generations, they can evolve

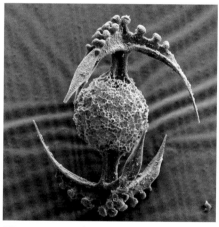

Siliceous plankton (radiolarian) from rocks in Turkey that are approx. 220 million years old (Triassic, Carnian). Photo taken by scanning electron microscopy. Size: approx. 0.4 mm at the largest dimension.

more rapidly. Over the course of hundreds of successive generations, mayflys are more likely to mutate in a decade than human beings in a thousand years.

Siliceous plankton (radiolarian) from rocks in Romania that are approx. 88 million years old (Cretaceous, Coniacian). Photo taken by scanning electron microscopy. Size: approx. 0.2 mm at the largest dimension.

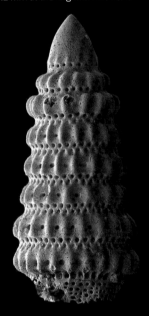

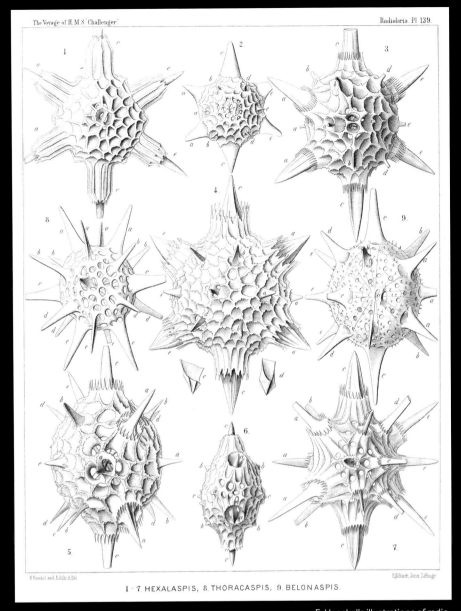

E. Haeckel's illustrations of radiolarians (siliceous plankton) from his report on the mission carried out by the ship H.M.S. *Challenger* (*Report on the Scientific Results of the Voyage of H.M.S.* Challenger *during the Years 1873–76,* volume XVIII, 1887).

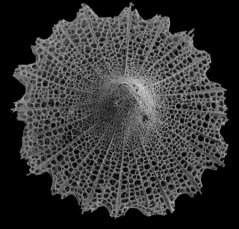

Siliceous plankton (radiolarian) from rocks in Turkey that are approx. 180 million years old (Jurassic, Pliensbachian). Photo, colorized, taken by scanning electron microscopy. Size: approx. 0.5 mm at the largest dimension.

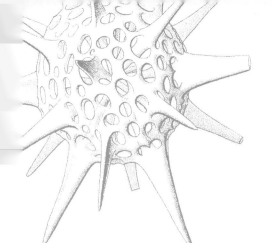

They Date the Opening
of the Ocean

Microfossils can be used to assign an age to the sediments and rocks in which they are found. This is often done without much notice as it usually seems rather trivial. From time to time, however, important events cannot be dated for lack of an effective indicator, and this prevents us from being able to reconstruct how these phenomena and their history unfolded. This was long the case for rocks that were found in the High Alps and that were previously at the bottom of an ocean 160 million years ago (during the Jurassic Period). These red rocks were formed from the accumulation of siliceous microplankton—radiolarians—and thus called radiolarites. Knowing their age was important because they sat on basalts and other rocks similar to those found on the parts of the ocean floor while it was in the process of opening and widening. In short, dating these rocks allowed us to know when this ocean finished opening. The problem was that their journey from where they were deposited in a sediment to their arrival at their current location on mountain summits was a difficult one, and the rocks had been transformed considerably. The embedded microfossils had suffered from these vicissitudes and were very poorly preserved; they had become unrecognizable. But one day perseverance was rewarded and a sample showed well-preserved radiolarians. It was possible to identify them and to give an age to the phenomenon. From this point forward, we could reconstruct the history of the beginning of the Alps! A few fragments of mere cubic millimeters allowed science to make a leap forward in its knowledge of the Earth's geologic history.

Top:
Siliceous plankton: radiolarian (*Perispyridium* sp.) from rocks in the Alps from approx. 160 million years ago. Photo taken by scanning electron microscopy. Size: approx. 0.2 mm at the largest dimension.

Below:
Siliceous plankton: radiolarian (*Parvicingula* sp.) from rocks in the Alps from approx. 160 million years ago. Photo taken by scanning electron microscopy. Size: approx. 0.2 mm at the largest dimension.

Large photo:
Lavas that erupt under water form sorts of balls, called "pillow lavas," between which fossil-rich sediments are sometimes trapped and thus allow lava flows to be dated.

Very Hard Rocks
from Fragile Objects

Some rocks achieve recognition by how they are used. This is especially true in the case of *pietra dura*, a technique that began to be used in 1588 by Grand Duke Ferdinand I de Medici for his marble marquetry projects.

Planktonic microorganisms, after their death, settle on the ocean floor. With time this ooze becomes rock, and this rock safeguards the memory of its initial components. Thus an ooze consisting of an accumulation of siliceous plankton and iron compounds will tend to take on a green or red color, depending on how oxidized the iron is. The rock that then forms from these very small elements will be thin, with a nice polish. In addition, the bright red color is used for decorative motifs.

This beautiful and very hard rock is not easy to carve. Moreover, it only exists naturally in relatively shallow shoals, measuring in the range of a few centimeters. It is therefore used to emphasize subjects and accentuate decorative elements. When it is bright red, this rock is called bloodstone or even martyr's stone because, according to mythology, this stone contains the blood of Christ.

Bloodstone has been used in very rare pieces, especially in the style of Florentine marquetry known as *pietra dura*. For a long time pieces of small stones from the Arno, the river that flows through Florence, were used to make these pieces, and then fine stones such as lapis lazuli, malachite and various kinds of jasper or agate were introduced. The use of colored marbles and organic materials, such as coral and mother-of-pearl, also became prevalent.

Different working techniques were developed over time, making it possible to create beautiful tableaus. The nuances of the colors and veins of the different stones are chosen with great care in order to obtain particular effects and create a realistic image. The colors of the stones give the appearance of shadows and thus three-dimensionality. The best example of this form of art is in the Medici Chapels at the Basilica of San Lorenzo in Florence.

Among the lavish pieces made for the Medici family were also marquetry tables that were offered as gifts to France.

Florentine marquetry depicting the city of Florence. The background of the plaque is white marble. The red flower consists of jasper: siliceous rock formed from radiolarian shells (siliceous plankton that lived approximately 150 million years ago).

Marquetry tabletop made in the Ligozzi style with semiprecious stones on a marble background. It would have belonged to Cardinal de Mazarin and successively to Colbert and Louis XIV. Italian marquetry (*pietra dura*) table from the 17th century. In the MNHN collection (French National Museum of Natural History).

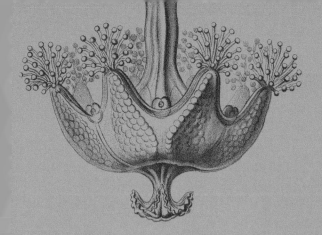

Architecture
and Nature

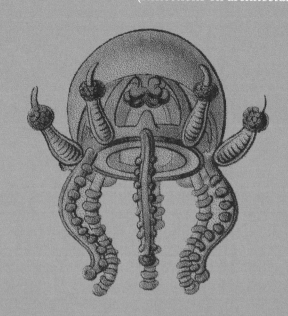

The beauty of forms found in nature was appreciated in the 19th century. Ernst Haeckel is the perfect example. For him, nature was closely related to art. His art was strongly influenced by the symmetry present in nature, especially that of microorganisms such as radiolarians (see "Haeckel: The Art of the Introduction," p. 54), and his drawings of planktonic organisms show this. His drawings were renowned. His scientific monographs were already successful, and his popular works titled *Kunstformen der Natur* (*Art Forms in Nature*), published between 1899 and 1904 in the form of numerous sets of prints, were found in all homes that aspired to be cultured.

Haeckel's representations of microorganisms, but also of macroorganisms and particularly of jellyfish, especially influenced the art of the early 20th century. The best examples of this fusion of art and nature can be seen at the Oceanographic Museum of Monaco. The jellyfish chandelier by Constant Roux, the four "radiolarium" lamps and the frescoes were all inspired by Ernst Haeckel's illustrations. Another example is the monumental door from the 1900 Paris Exposition Universelle made by the French architect

Art Nouveau glass "Jellyfish" chandelier made by Constant Roux for the museum in Monaco. Inspired by plates by E. Haeckel.

René Binet. Binet's publication *Esquisses décoratives* (*Decorative Sketches*), inspired by Haeckel, was one of the foundations of Art Nouveau. After Binet came another French architect, Robert Le Ricolais, who would follow this same path in the 1930s.

Plates of E. Haeckel's illustrations of jellyfish from his published work *Kunstformen der Natur* (*Art Forms in Nature*), 1899–1904.

Art Nouveau living room chandelier, of which the resemblance to the figure in the Haeckel illustration below can be seen.

Art Nouveau chandelier inspired by jellyfish at Hotel Negresco in Nice, France.

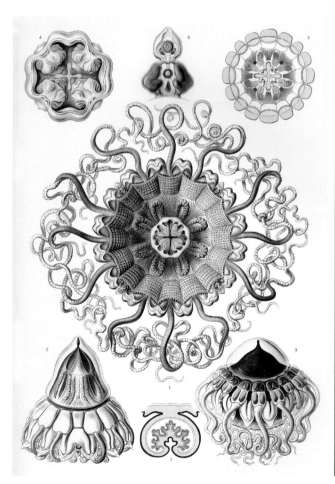

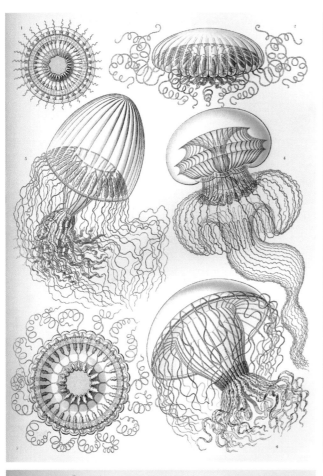

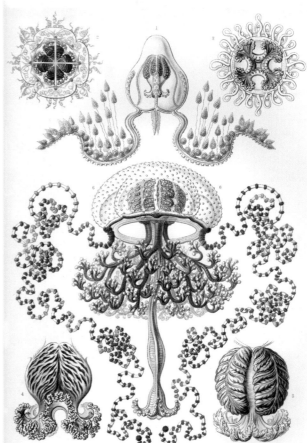

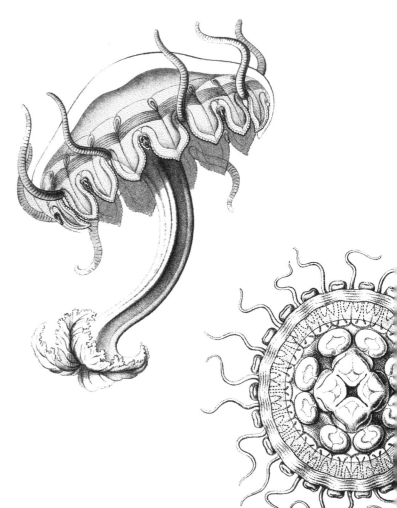

Nature
according to Le Corbusier

A work of architecture is the result of arranging different elements that build unity. During the 18th and 19th centuries, this type of interdependency was likened to certain natural organisms.

The propagation of the idea of evolution inspired the notion that architectural styles could also "evolve." Biologists increasingly considered natural forms to be the result of both their aptitude for a specific function and morphological norms. This school of thought culminated with the publication in 1917 of D'Arcy Wentworth Thompson's book *On Growth and Form*, which was a huge success among architects.

The functionalist school of thought, which believed the form and external appearance of a building must flow from its function, was spread by the Austrian biologist Raoul Francé in the 1920s. In a work published in 1923, Francé wrote that "necessity prescribes certain forms for certain qualities. Therefore, it is always possible . . . to infer the activity from the shape, the purpose from the structure. In nature, all forms . . . are a creation of necessity." The parallels with architecture are obvious, and the illustrations that Francé pro-

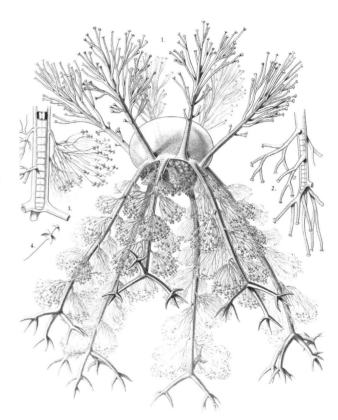

vided were adopted in works by leaders of the modernist movement. Foremost among them was Le Corbusier, who stated, "Biology is henceforth the key word in architecture and town planning."

Radiolarians drawn by E. Haeckel, following the *Challenger* expeditions, in his work published in 1887 (*Report on the Scientific Results of the Voyage of H.M.S.* Challenger *during the Years 1873–76*).

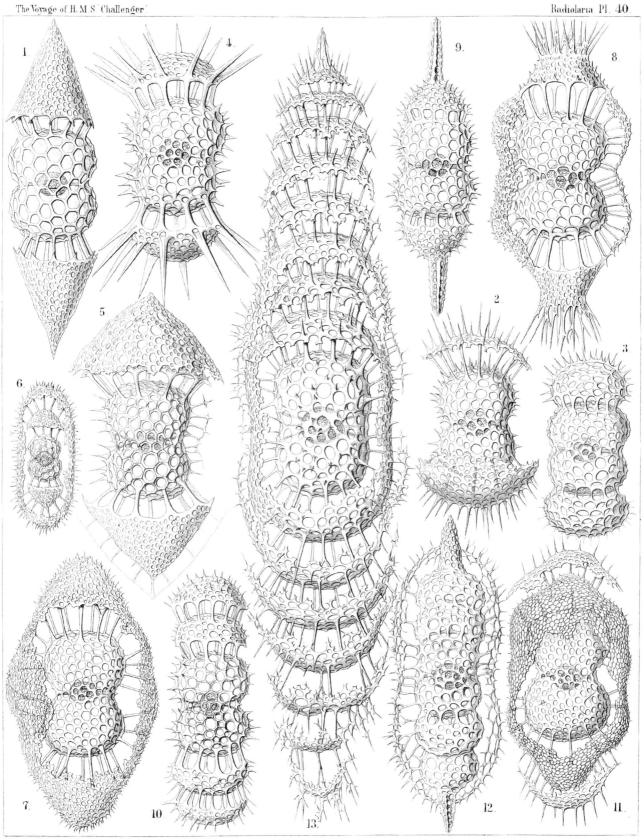

213

E.Haeckel and A.Giltsch Del.

C.Giltsch, Jena, Lithogr.

1–3. PANARTUS, 4. PANARTIDIUM, 5–8. PERIPANARTUS,
9. PANARIUM, 10. OMMATOCAMPE, 11–13. ZYGARTUS.

Art Nouveau
and the Microscopic World

Art Nouveau was an artistic movement of the late 19th and early 20th centuries based on the aesthetics of curved lines, colors and ornamentation inspired by nature. It was not only concerned with embellishment but also with structure. It would later evolve into Art Deco. Born in reaction to the abuse of industrialization and conventional large-scale reproduction, it quickly grew into an international movement. By being one of the first to draw a multitude of shells (microscopic or not), flowers and jellyfish for scientific purposes, Ernst Haeckel is considered a herald of Art Nouveau.

In France, Eugène Viollet-le-Duc did not stop using the current material of the times (namely, iron). Quite the opposite, he wanted to display it by giving it an ornamental and aesthetic function, like in the Gothic structures of the Middle Ages. Viollet-le-Duc would inspire many Art Nouveau architects. In 1893, Victor Horta built the Hôtel Tassel in Brussels, perfectly in line with Eugène Viollet-le-Duc's style. These creators were quickly overtaken by the success of the trend they inspired, which triumphed at the 1900 Paris Exposition Universelle. Art Nouveau sought to achieve unity of art and life. Choosing nature as a

Gaudí's Sagrada Família in Barcelona (Spain) seen from Casa Milà and inspired by forms found in nature.

source of inspiration was a reaction against the rationalism of the beginning of the industrial era and its cold efficiency. Barcelona is home to some Art Nouveau monuments, notably architect Antoni Gaudí's famous Basilica of the Sagrada Família. Its construction began in 1882 and is still not complete.

The Passion facade of the Sagrada Família appears to be inspired by radiolarians.

Detail of stained glass windows seen from inside that clearly evoke the pore arrangements of radiolarians.

Main cupola inside the Sagrada Família that leans on its pillars, like the segments of radiolarians that extend spines.

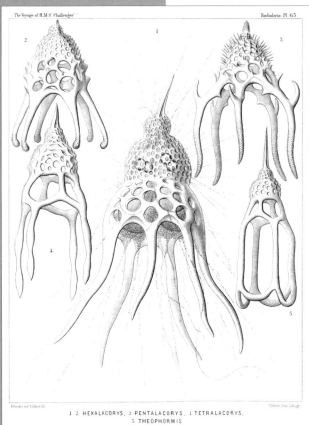

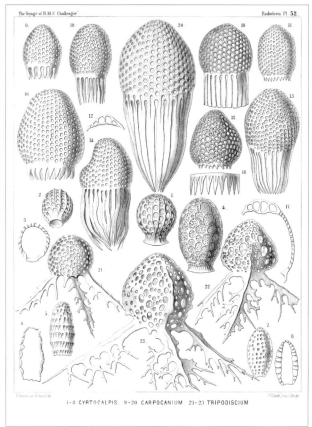

Illustrations of siliceous planktonic organisms (radiolarians) by E. Haeckel in his work on the forms collected during the H.M.S. *Challenger* expedition (*Report on the Scientific Results of the Voyage of H.M.S.* Challenger *during the Years 1873–76*, volume XVIII, 1887).

A Plea to Be Inspired by Nature's "Inexhaustible Fountain of Beauty"

It is such an insect that neither by day nor night, neither to the naked eye nor under the microscope, could excite a feeling of interest; but if you go to the effort of lifting up, with a delicate and patient scalpel, the layers that compose the thickness of its scaly wing, you will find there, in most instances, a variety of unexpected designs, sometimes plant-like curves, sometimes airy branches, sometimes angular striated figures, like hieroglyphics, which remind you of the letters of certain Oriental languages . . . Frankly, what is equal, or approaching from afar, in our arts? How much they would need, in their seemingly fatigued and languid condition, to be revived by these living sources! In general, instead of going straight to Nature, to the inexhaustible fountain of beauty and invention, they have solicited help from scholarship, the arts of times gone by, and the history of man.

. . . Our intelligent Parisian merchants, who have reluctantly followed the path laid out for them by the great producers, may one day escape from these rich and heavy styles. Someone will lose patience, and, turning his back on the copyists of old things, will seek advice from Nature herself, from the great collections of insects, from the greenhouses of the Jardin des Plantes.

Michelet, Jules. "De la rénovation de nos arts par l'étude de l'insecte." In *L'Insecte.* ["On the Renovation of Our Arts by the Study of the Insect." In *The Insect.*] Paris: Librairie Hachette et Cie, 1858.

Illustrations of radiolarians (siliceous plankton) published by E. Haeckel in his report on the mission carried out by the ship H.M.S. *Challenger* (*Report on the Scientific Results of the Voyage of H.M.S.* Challenger *during the Years 1873–76*, volume XVIII, 1887).

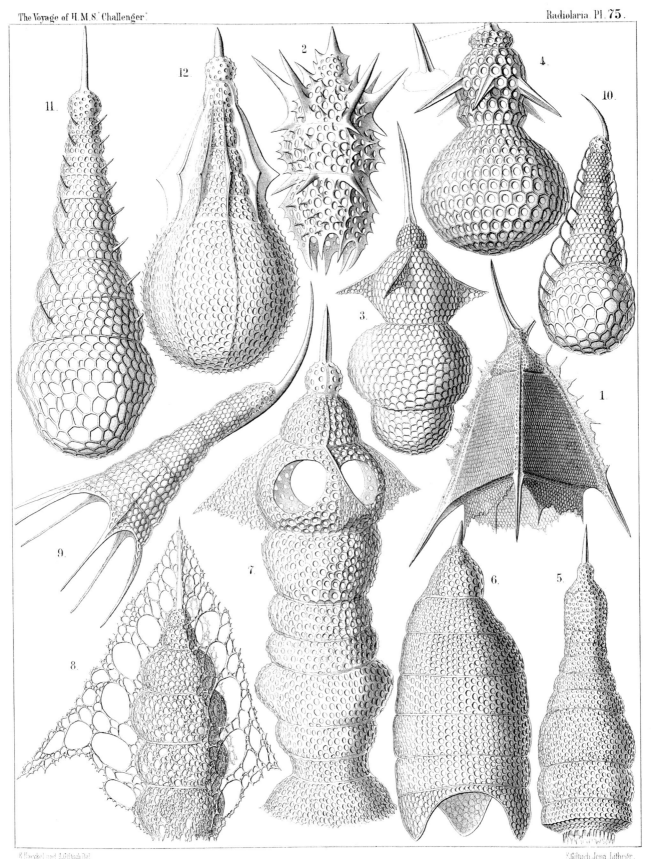

1. ARTOPILIUM, 2. ARTOPHORMIS, 3. ARTOPERA, 4. ARTOPHATNA, 5. STICHOCORYS,
6. STICHOPODIUM, 7. CLATHROPYRGUS, 8. STICHOPTERYGIUM, 9. STICHOPHORMIS,
10. CYRTOLAGENA, 11. STICHOPERA, 12. STICHOPHATNA.

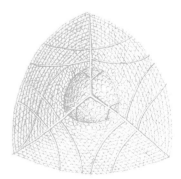

*What we call chance
is only our inability to understand
a higher degree of order.*

Jean Guitton

Varied Yet Constrained
Forms

Micropaleontologists study only skeletons. These are interchangeably called skeletons, shells and tests.

A radiolarian skeleton is built according to a well-defined geometric plan. It is composed of one or more siliceous shells (spherical, concentric, conical, etc.) from which spines eventually radiate. Its architecture can be very simple (a single spicule) or very complex.

The geometry of the skeleton responds to the same laws of physics (surface tensions) as those that govern the interactions between fluids or between fluids and solids. This skeletal geometry can therefore be compared to different physical models that apply to structural features. It is no coincidence that we find a hexagonal structure for flat structures, such as in foams, honeycomb and radiolarian shells, whereas pentagons are inserted for more spherical forms, like soccer balls. There is a striking similarity between the forms of what appears to be an unruly group of soap bubbles and some skeletons. Similarly, soap bubbles of different sizes group together according to a model found in nature, either rectilinearly or curvilinearly.

It is always striking to observe this sort of magic: when we remove a curvilinear tetrahedron made of iron wire from a soapy bath, the form is perfectly identical to that of certain radiolarians (*Callimitra* sp.). Magic is simply the result of physical forces that behave similarly.

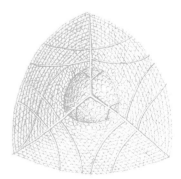

Illustrations of radiolarians
(siliceous plankton)
published by E. Haeckel in
his report on the mission
carried out by the ship H.M.S.
Challenger (*Report on the
Scientific Results of the
Voyage of H.M.S.* Challenger
during the Years 1873–76,
volume XVIII, 1887).

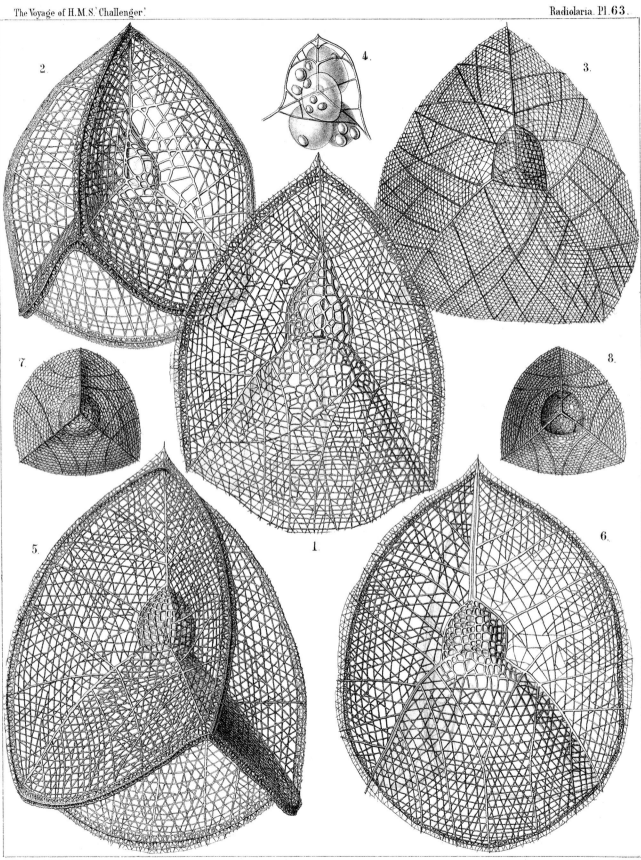

CALLIMITRA.

219

Architects, Creators and Markers of Time

René Binet's
Radiolarians

René Binet (1866–1911) was an architect and painter best known for being the creator of the monumental gate at the Place de la Concorde made for the 1900 Exposition Universelle in Paris and Le Printemps store. For the gate, Binet wanted to achieve something "that has never been done in architecture, an architecture of color and light."

Binet was influenced by his memories of polychrome architecture in Venice, by *Theory of Colours* by the German poet Johann Wolfgang von Goethe (1749–1832) and by Ernst Haeckel's biological works, which he consulted extensively at the library of the National Museum of Natural History in Paris.

A practitioner of Art Nouveau, he made numerous sketches of buildings, furniture and decorative objects, some of which were published in 1905 with a long, detailed preface by the journalist and art critic Gustave Geffroy (1855–1926) titled "Préface aux *Esquisses décoratives* de Binet, architecte" ("Preface to *Decorative Sketches* by Binet, architect"). According to Geffroy, the monumental gate at the Exposition Universelle was inspired by a group of radiolarians called Cyrtoidea and by *Pterocanium trilobum* in particular. He specified that Binet considered the radiolarian *Clathrocanium reginae* to be the most beautiful creature.

Top:
The monumental gate by René Binet. Entrance to the Paris Exposition Universelle, 1900, inspired by a radiolarian (*Pterocanium trilobum*).

Bottom:
Illustrations by E. Haeckel: some radiolarians (Cyrtoidea) that the architect Binet preferred. He considered *Clathrocanium reginae* (top row, second from the left) to be the most beautiful, and *Pterocanium trilobum* (rightmost in the middle row) is the one that inspired the monumental gate.

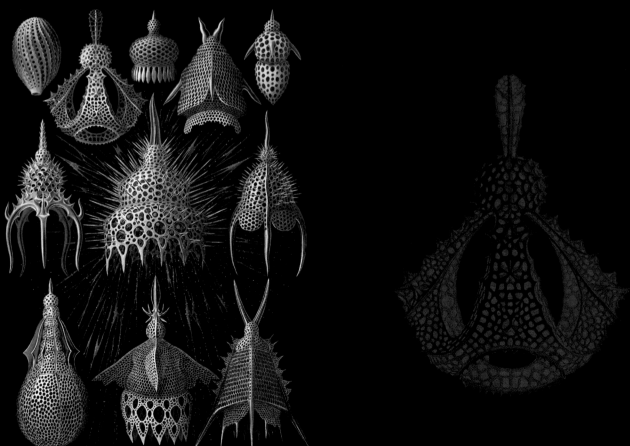

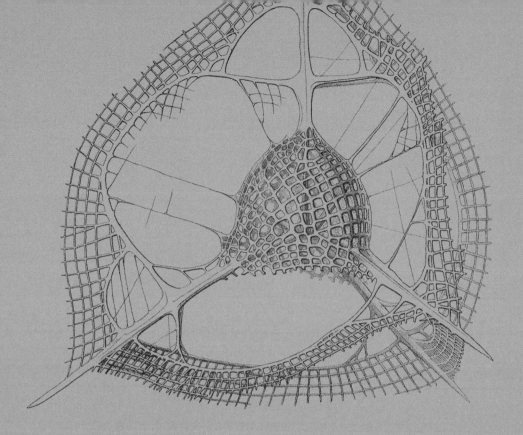

Protists:
Animals and Plants

In the 16th lesson of his History of Creation, *Haeckel says the following: "We see perfectly well that beings are divided into plants and animals, and it does not occur to us that these beings might stem from one common source. Yet it is true, or at least seems to be, since as we climb the ladder of beings we arrive, beyond ferns and mollusks, at a group, called the protists, of which most of the representatives are so small in volume that they are barely visible or are invisible to the naked eye. The game of their lives is such a singular mixture of animal and plant properties that we are justified in not classifying them in any kingdom. The contradictions that have arisen about them are not due to the imperfection of our knowledge about protists, but rather to their unicellular nature."*

Gustave Geffroy. Preface to *Esquisses Décoratives* [*Decorative Sketches*]. By René Binet. Paris: Librairie Centrale des Beaux-Arts, 1905.

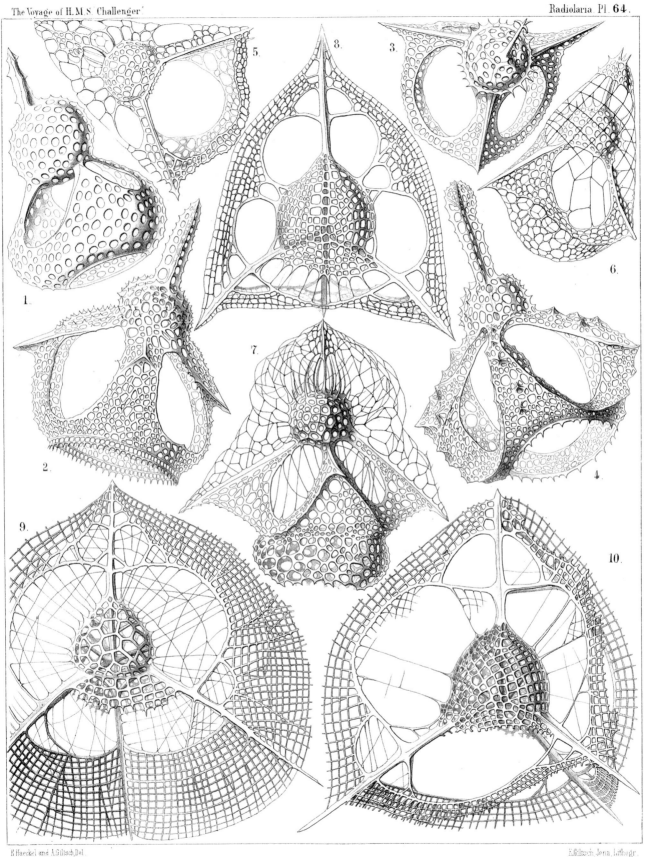

E.Haeckel and A.Giltsch,Del.

E.Giltsch Jena,Lithogr.

1-4. CLATHROCANIUM, 5-7. CLATHROLYCHNUS, 8-10. CLATHROCORYS.

Micropaleontologists
Called Upon for Art Nouveau

This means of addressing the large laboratory of Nature, always in motion, always in production, without a moment's pause or hesitation. There, one can obtain the infallible secret of creations and transformations. . . .

To his credit, Mr. René Binet had a sudden willingness. He went to the invisible world, to the infinite forms first revealed by the microscope, and he studied, with passionate attention, the general characteristics of these forms and the abundance of their by-products, and he learned about the perpetually renewed life that hides in the depths of the ocean, all this universe in development from which the separate forms are constantly emerging from the transient mixture of mineral life, vegetable life and animal life. . . .

And here is our artist, following the learned professor of the University of Jena [Ernst Haeckel], who was preceded in this study by [Christian] Ehrenberg and [Alcide] d'Orbigny, who examined the major divisions of this intermediary kingdom of protists . . .

From then on, it became possible for Mr. René Binet to draw inspiration from this information for a purely geometric decorative element, such as [a] tile . . . derived from a . . . diatom . . . divided into six sectors: three black pieces adorned with six white dots, surrounding a sort of larger rose, and three gray pieces adorned with very fine points, enhanced by black dots. This opposition naturally created two tones in the resulting tiles: red and black, for example.

Gustave Geffroy. Preface to *Esquisses Décoratives* [*Decorative Sketches*]. By René Binet. Paris: Librairie Centrale des Beaux-Arts, 1905.

Illustrations of diatoms (siliceous plankton) by E. Haeckel from his published work *Kunstformen der Natur* (*Art Forms in Nature*), 1899–1904.

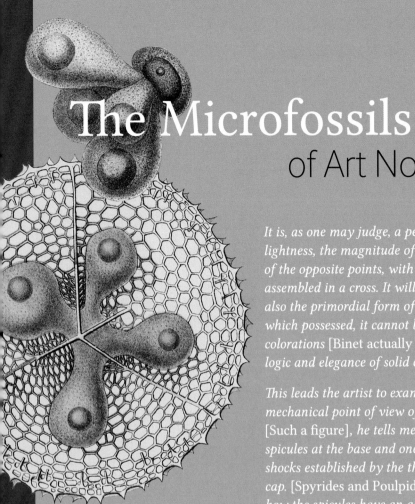

The Microfossils
of Art Nouveau

It is, as one may judge, a perfect harmony of curves, an extraordinary lightness, the magnitude of points that support the effort, a throttling of the opposite points, with a marvelous crowning of four spearheads assembled in a cross. It will be a tiara, a diadem if you will. It is also the primordial form of the Monumental Gate of the Exposition, which possessed, it cannot be denied, not only the charm of luminous colorations [Binet actually wanted to play with electricity], *but the logic and elegance of solid construction.*

This leads the artist to examine Radiolarians from the purely mechanical point of view of the resistance of their arches and walls. [Such a figure], *he tells me,—formed of a half-sphere, of three spicules at the base and one at the top, has its resistance against shocks established by the three perforated wings that straddle the cap.* [Spyrides and Poulpidae] *By this schema . . . , we see very well how the spicules have an absolute connection with the perforated sphere whose center is at the junction of the four spicules, the hollow sphere being as resistant as if it were full, which immobilizes the four edges incorporated in its volume. The same observation is applicable* [a different figure], *to a volume that keeps its shape through the placement of mesh between the spicules,—*[a different figure], *with the two axes intersecting in the same plane of unequal length, and so to speak blocked in a perforated ellipsoid,—*[a different figure], *more complete, since its three axes intersect as well as the axes of symmetry of the cube.*

Gustave Geffroy. Preface to *Esquisses Décoratives* [*Decorative Sketches*]. By René Binet. Paris: Librairie Centrale des Beaux-Arts, 1905.

Illustrations of radiolarians (siliceous plankton) published by E. Haeckel in his report on the mission carried out by the ship H.M.S. *Challenger* (*Report on the Scientific Results of the Voyage of H.M.S.* Challenger *during the Years 1873–76,* volume XVIII, 1887).

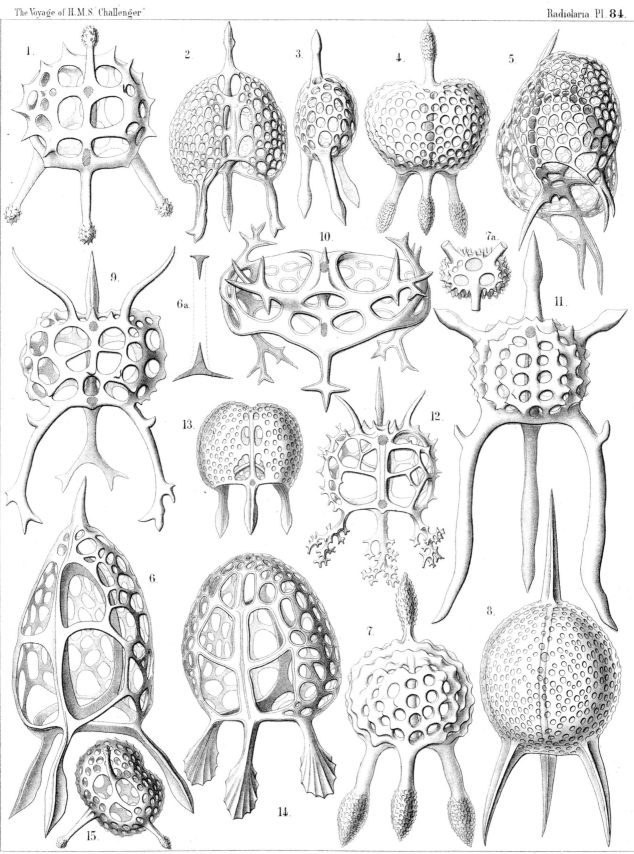

1–8. TRIPODOSPYRIS, 9–12. TRICERASPYRIS, 13–15. TRISTYLOSPYRIS.

227

Architects, Creators and Markers of Time

An Economy
of Material

We saw in "The Economies of Energy" (p. 56) that this advantage of efficient living things led to light and perforated forms. This observation had already been made in the 19th century by art critics:

The more we look at these figures, the better we perceive that a law of mechanics, which is a vital law, has prescribed in each of these volumes the suppression of all useless material, leaving only the edges and surfaces which must bear stress, and added to them, in some cases, the elements likely to ensure the rigidity of the whole. Nothing superfluous. This is the important lesson.

Such as it appears to us, in its hidden splendor, this world, tiny, invisible. When it is under the light, and the microscope reveals it, it flourishes into thousands and thousands of fantasies, based on reason, keeping the unexpectedness of a sketch, the charm of spontaneity. Strange perforated cupolas, light arcades, all these slight things are so well proportioned that they can lend themselves to all developments, reaching colossal proportions. There are building plans for sheltering mankind inside the slightest atom of the animated mud that is piling up at the bottom of the seas.

Gustave Geffroy. Preface to *Esquisses Décoratives* [*Decorative Sketches*]. By René Binet. Paris: Librairie Centrale des Beaux-Arts, 1905.

Radiolarians develop a whole panoply of spines that allow the organism to lengthen the large extensions of its cytoplasm and thus expand its field of investigation to find prey. It only appears to have grown in size when really this ingenuity does not require much extra material for the construction of its skeleton.

Radiolarians have very varied forms, but not just any forms. Rather, they are more like variations on a theme with a certain number of strict constraints that nevertheless allow for a variety of possibilities. A nice lesson from nature.

Illustrations of radiolarians (siliceous plankton) published by E. Haeckel in his report on the mission carried out by the ship H.M.S. *Challenger* (*Report on the Scientific Results of the Voyage of H.M.S.* Challenger *during the Years 1873–76*, volume XVIII, 1887).

229

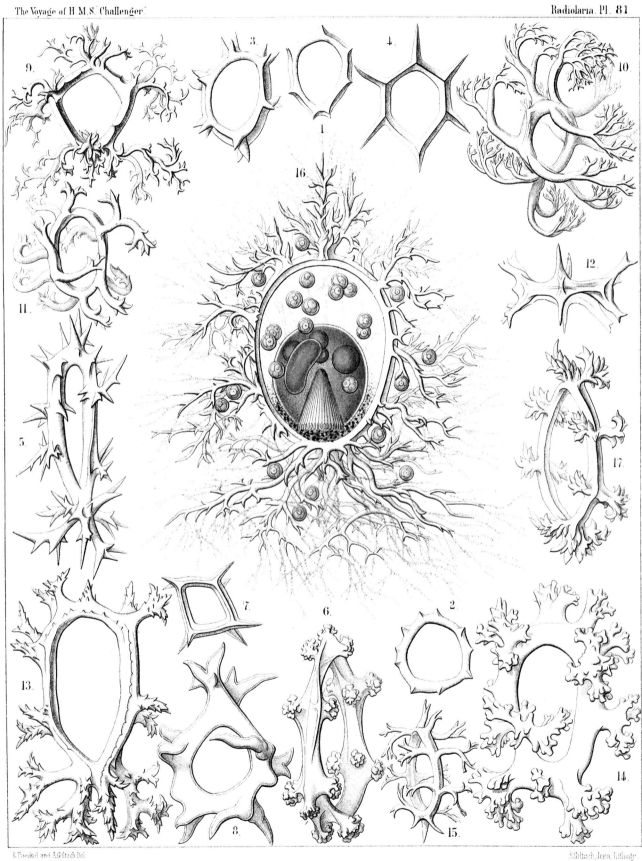

G.Haeckel and A.Giltsch Del

A.Giltsch, Jena, lithogr.

1 - 8 LITHOCIRCUS, 9 - 17 DENDROCIRCUS.

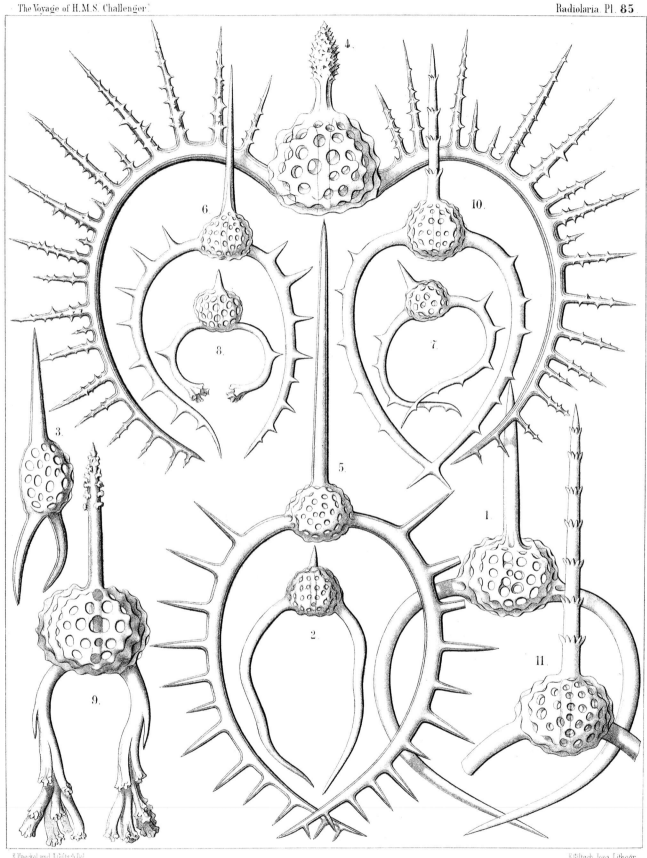

E.Haeckel and A.Giltsch Del.

E.Giltsch, Jena, Lithogr.

1 – 3. DIPODOSPYRIS, 4 – 11. DORCADOSPYRIS.

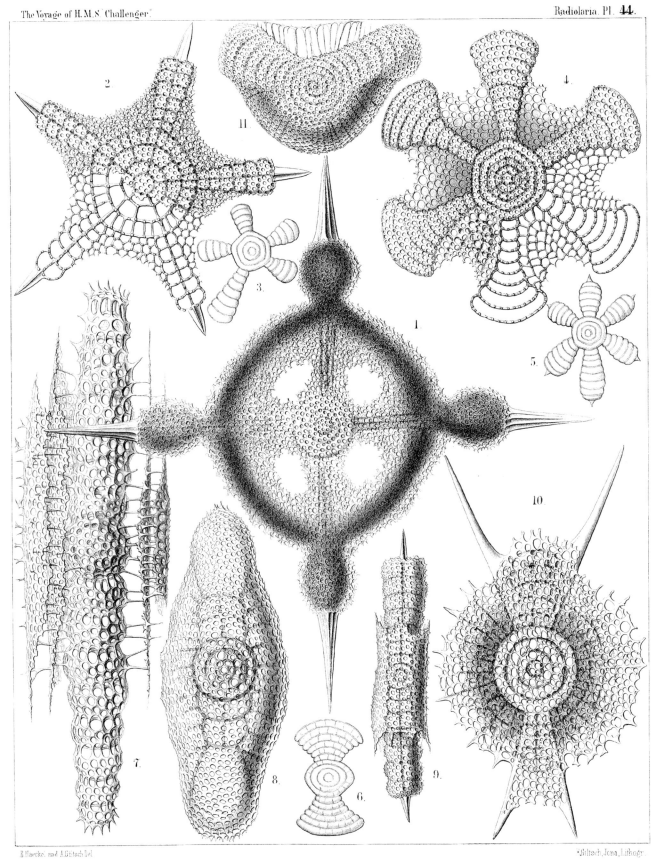

E.Haeckel. and A.Giltsch.Del.   Giltsch, Jena, Lithogr.

1 STEPHANASTRUM, 2.3.PENTALASTRUM, 4.5.HEXALASTRUM,
6.AMPHIBRACHIUM, 7-11 AMPHYMENIUM.

*The architect of the future will build by imitating nature,
because it is the most rational,
long-lasting and economical of methods.*

Antoni Gaudí

Nature
Inspires Architecture

*And for the particular application of such research to architecture and
the forms of objects, it seems to me that it justifies itself through these
examples. In short, Mr. René Binet searched the invisible world for what
others have looked for among flora.*

*But the flower, supple and malleable, offers itself for ornamentation,
whereas all this invisible world teems with rigid, defined and complete
forms, all ready for architecture. It is there, at the point where science
makes us witness one of the states of the evolution of species, where it
catches by surprise the suspected unity of the material, it is there that
this artist, this very humble and very modest schoolboy, has sought
to collect the lessons of forms and movements that could give him the
world of hatching and growing things. It is there that he has taken
all these lines, all these angles, all these circles, all these ellipses, all
these stars, all these figures that become, in his pencil's path, a sort of
extraordinary living geometry.*

*His imagination has done more, and his personal contribution is not
only a choice of means and materials, something that would already
be a considerable and infinitely useful undertaking. By an operation of
mind that seems quite simple now that it has been followed by exe-
cution, but that was necessary to conceive and carry out successfully,
Mr. René Binet saw, in this invisible and precise life toward which he
leaned with a desire so eager to penetrate and assimilate, not only the
means of creating architectural forms and sculptural ornamentations of
objects, but also, magically, the objects themselves, ready-made, ready
to be realized.*

Gustave Geffroy. Preface to *Esquisses Décoratives* [*Decorative
Sketches*]. By René Binet. Paris: Librairie Centrale des Beaux-Arts,
1905.

Shapes so varied that they are inspirational for
architecture. Illustrations of radiolarians (siliceous
plankton) published by E. Haeckel in his report
on the mission carried out by the ship H.M.S.
Challenger (*Report on the Scientific Results of
the Voyage of H.M.S.* Challenger *during the Years
1873–76*, volume XVIII, 1887).

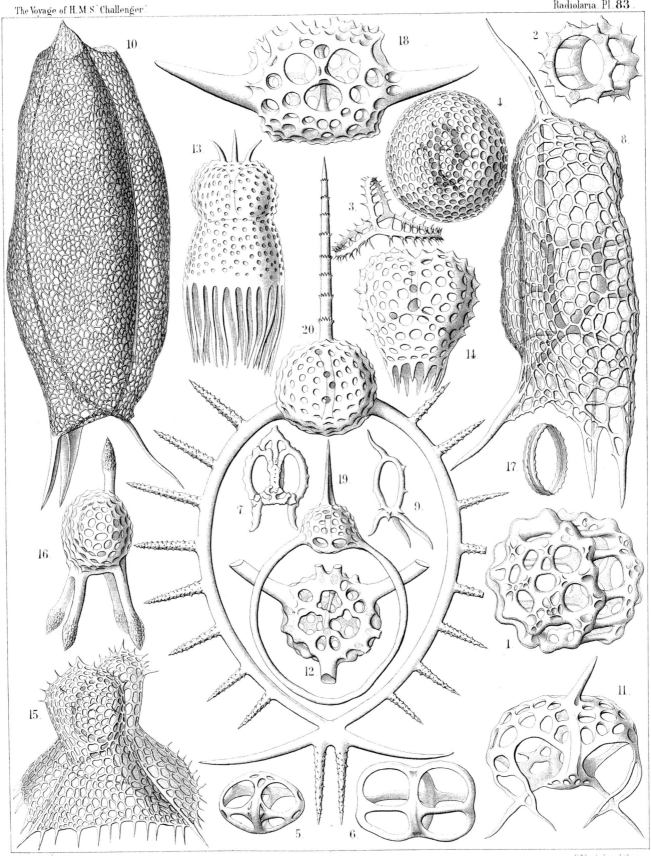

1.2. LITHOTYMPANIUM, 3. DYOSTEPHANUS, 4. SPHAEROCIRCUS, 5.6. TRISSOCYCLUS,
7. DIPOCORONIS, 8 – 10. LAMPROSPYRIS, 11.12. CLADOSPYRIS, 13. RHODOSPYRIS,
14.15. DESMOSPYRIS, 16.17. TETRASPYRIS, 18–20. STEPHANOSPYRIS.

Praise for Scientists
from an Art Critic

With him, I saw the admirable Radiolarians, the most perfect class of the Rhizopoda group. Thanks to the kindness of the learned professor of the French National Agricultural Institute, Mr. Lucien Cayeux, we were able to admire the most perfect specimen of the group under the microscope. It was especially in the wake of the beautiful English expedition aboard the Challenger, *led by the zoologist Wyville Thomson, that the Radiolarians were discovered and studied, and it is thanks to the research done on these animals, almost unknown until then, by Professor Haeckel, that we can glimpse the infinite richness of the protist kingdom. But I must return to the forty-six volumes published on the* Challenger *expedition, in which the plates are accompanied by purely naturalistic explanations. No scientist thought it necessary to leave his domain in order to show the mathematical or static sides of these forms, just as no artist dreamed of entering this living world to capture the strength of construction that supports it or the grace of ornamentation that adorns it.*

Gustave Geffroy. Preface to *Esquisses Décoratives* [*Decorative Sketches*]. By René Binet. Paris: Librairie Centrale des Beaux-Arts, 1905.

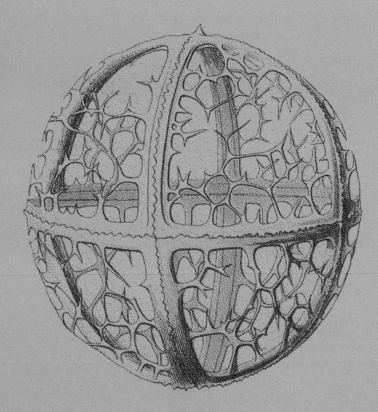

Illustrations of radiolarians (siliceous plankton) published by E. Haeckel in his report on the mission carried out by the ship H.M.S. *Challenger* (*Report on the Scientific Results of the Voyage of H.M.S.* Challenger *during the Years 1873–76*, volume XVIII, 1887).

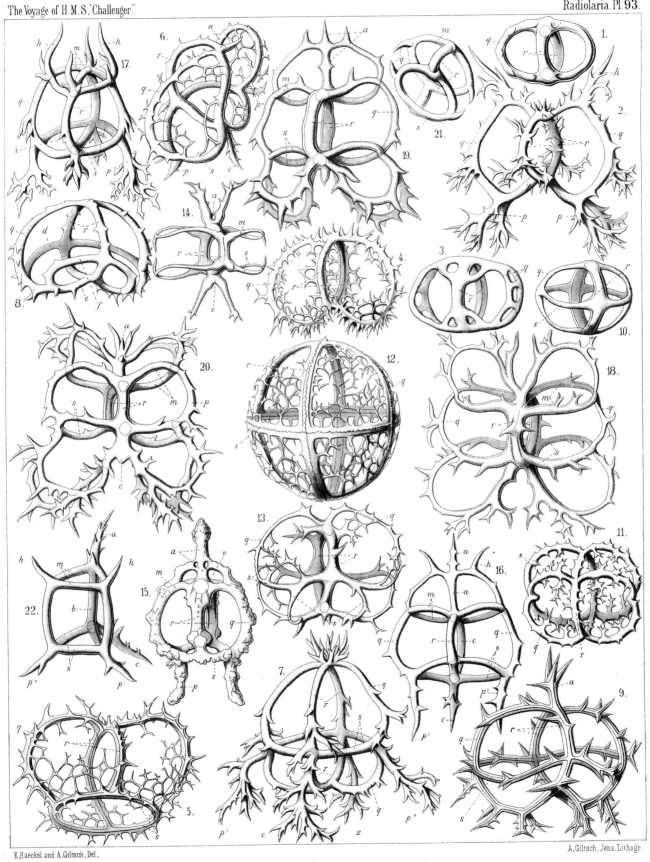

E.Haeckel and A.Giltsch, Del.

A.Giltsch, Jena, Lithogr.

1-4. ZYGOSTEPHANUS, 5-6. ACANTHODESMIA, 7-13. TRISTEPHANIUM,
14-17. ACROCUBUS, 18-20. TOXARIUM, 21.22. PRISMATIUM.

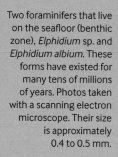

Foraminifers:
Indicators of Time and Environment

Foraminifers live in the water. Some planktonic varieties in the ocean float in the water column, especially in the upper part that is exposed to light, but sometimes they are found in the ocean's dark abysses. Others live on the seafloor, or in the sediment; these are benthic. Foraminifers are found in all types of salt water and sometimes even in fresh water. They are found in all the seas of the world. Some species, called burrowers, feed by digging in the shallow sediment. Planktonic species are equipped with floats, in the form of pockets of gas or oil; they are generally round (like *Globigerina*) and sometimes have thorny structures, especially those living near the surface. Appearing in the Cambrian Period (about 500 million years ago), foraminifers especially began developing in the Triassic Period (about 250 million years ago). Since then, they have sometimes been so abundant that they constitute a major component of the marine sediment. Their size usually ranges from 0.03 to 1 millimeter (some may be more than 18 cm). Continuously evolving over time, foraminifers are very useful tools for dating rocks.

Some foraminifer forms favor particular environments, meaning that they are also good markers of past environments. For example, when the Rance Tidal Power Station in France began operating, it disturbed the local environment enough that foraminifers developed morphological anomalies (mutations). For the same reasons, foraminiferal fauna with many anomalies appeared in some bays affected by volcanic activity in Japan. Observing changes in foraminifers is also a tool for identifying pollution in certain areas that would have gone unnoticed otherwise.

Two foraminifers that live on the seafloor (benthic zone), *Elphidium* sp. and *Elphidium albium*. These forms have existed for many tens of millions of years. Photos taken with a scanning electron microscope. Their size is approximately 0.4 to 0.5 mm.

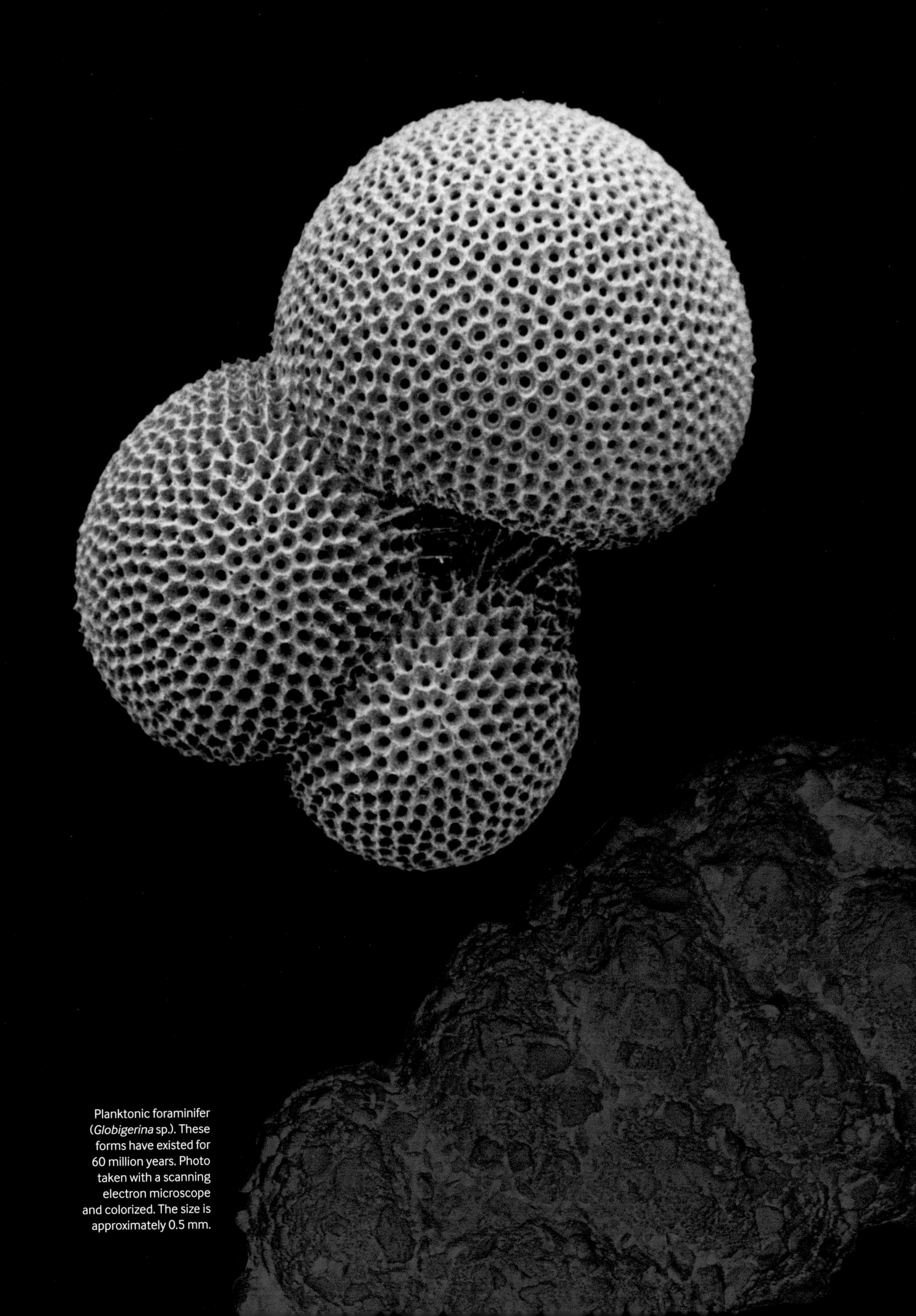

Planktonic foraminifer
(*Globigerina* sp.). These
forms have existed for
60 million years. Photo
taken with a scanning
electron microscope
and colorized. The size is
approximately 0.5 mm.

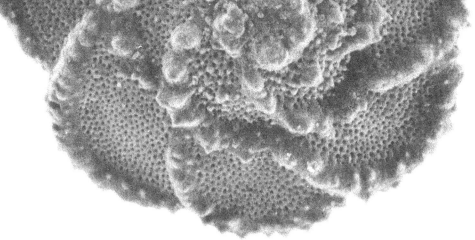

*For some, who are travelers,
the stars are guides.
For others they are no more than
little lights in the sky.
For others, who are scholars,
they are problems.*

Antoine de Saint-Exupéry,
Le Petit Prince
(*The Little Prince*)

A Single Microfossil:
A Guideline for the TGV

After numerous geotechnical studies carried out for the Channel Tunnel, it was decided to construct it by tunneling through the layer of chalk called "chalk marl" because of its plasticity (therefore making it less sensitive to fracturing) and, above all, because its high clay content made it waterproof. The boundary between the chalk marl and the gray chalk that sits on top of it is not detectable by physical methods because there is an insufficient difference in their densities. Furthermore, the difference between these two chalks is difficult to discern with the naked eye through surveying. Therefore, for the Channel Tunnel it was necessary to find reliable markers that could quickly identify the most effective tool. Among the many fossils contained in chalk, some have only existed for a very brief amount of geologic time (in the range of 50,000 years). This is true of the foraminifer *Rotalipora reicheli*, which occurs at the top of the chalk marl layer, thus determining the location of the top surface at a point in time.

Even with the most advanced technologies of the 1990s, it was a microfossil that had lived some hundred million years ago (during the Cenomanian) that we relied on to execute this civil engineering project we had been eagerly awaiting for two centuries.

Dating rocks by identifying the microfossils they contain has many applications, including geotechnical studies for the construction of dams, prospecting for mining (surveys for the study of deposits) and so on.

Foraminifer (*Rotalipora reicheli*). This microfossil served as a guideline for the tunnel boring machine under the English Channel. Its largest diameter reaches 0.5 mm.

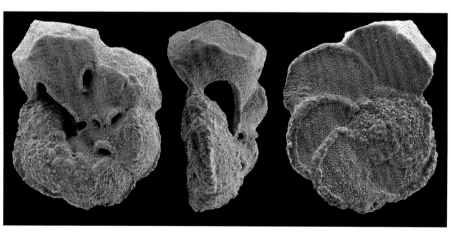

The TGV Eurostar
passes through the
Channel Tunnel,
which was dug using
a microfossil as a
guideline.

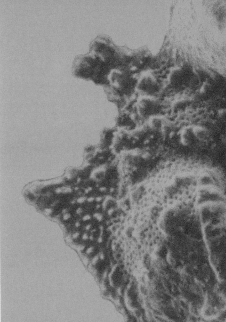

*Wherever anything lives,
there is, open somewhere,
a register in which time is
being inscribed.*

Henri Bergson

Microfossils:
Timekeepers of Industry

In oil exploration, one of the fundamental challenges is knowing the arrangement of the layers of the subsoil so that the "petroleum system" can be determined. In other words, the challenge is being able to identify and locate the rocks where the oil was formed (source rocks), the rocks where oil has been stored (reservoir rocks) and the rocks that prevent oil from migrating farther (caprocks). Only then can we locate deposits.

Fossils are the best tool for meeting the demands of cost and speed. Evidently, the fossils visible to the naked eye within the rocks traversed by a drill pipe will generally be crushed by the drill bit and become hardly recognizable. Microfossils, on the other hand, are of course subjected to the same rock crumbling, but they are so small that they can be found in large quantities in the chips (rock debris) that result from the drill bit fragmenting the rock, and they remain perfectly identifiable. They thus provide information on the age of the rock and, potentially, on the environmental conditions during its formation. This data is useful, even necessary, for petroleum geologists. Therefore, microfossils make it possible to prevent ongoing core drilling, which is very expensive.

Oil drilling rig
in Siberia.

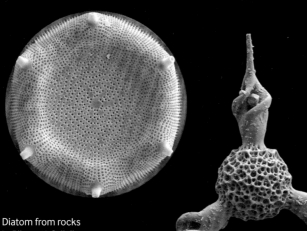

Diatom from rocks in Păușești-Otăsău (Romania) from the Miocene (Badenian Stage, 12 million years ago). External view, electron microscopy.

Radiolarians (*Capnuchospharea triassica*) from rocks in Turkey (Elbistan) from the Triassic Period (Carnian, 230 million years ago). Photos taken with a scanning electron microscope. Size: approx. 0.5 mm.

Diatom (*Entogonia formosa*) from rocks in Păușești-Otăsău (Romania) from the Miocene (Badenian Stage, 12 million years ago). External view, electron microscopy.

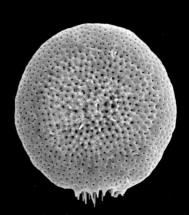

Radiolarian (*Spongodiscus* sp.) from rocks in Păușești-Otăsău (Romania) from the Miocene (Badenian Stage, 12 million years ago). This species sometimes has a sheath covering the central part. Scanning electron microscopy.

Radiolarian (*Hexacontium* sp.) from rocks in Păușești-Otăsău (Romania) from the Miocene (Badenian Stage, 12 million years ago). This specimen is a bit of a mutant, in the biological sense of the word, because, as its name indicates, the species usually has six spines and this one has seven. Scanning electron microscopy.

Radiolarian (*Dictyocoryne* sp.) from rocks in Păușești-Otăsău (Romania) from the Miocene (Badenian Stage, 12 million years ago). Scanning electron microscopy.

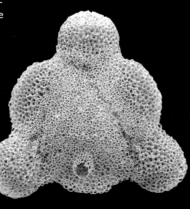

Radiolarian (*Spongotortilis* sp.) from rocks in Turkey (Elbistan) from the Triassic Period (Carnian, 230 million years ago). Photos taken with a scanning electron microscope. Size: approx. 0.3 mm.

Microfossils
for Land Use Planning

Land use planning is done based on need and technical possibilities. In this respect, the type and structure of rocks are considerable constraints. Indeed, if we look at the establishment of certain transportation routes, for example, there seems to be a correlation with the type of rock visible on the surface. The preference for the TGV Paris-Lille line was to build it on chalk rather than on clay, as clay is much less stable from a geotechnical point of view. The same was true for the highways that connect Lille to Mons (Belgium) and Lille to Calais. An example that shows the significance of the choice to build on chalk is demonstrated by the highway leading from Lille to Dunkirk. It was built on clay that lacked a chalky substrate, so it is often damaged and requires frequent work.

When rocks are less easy to differentiate, geologists use information provided by microfossils. In addition to their chronological contribution, microfossils make it possible to ensure proper monitoring of sedimentary layers and, consequently, to identify fractures such as faults. This is the reason why the geological and structural maps of the chalk in eastern France were created based on very detailed meshing of paleontological surface and survey data. These maps provide the basis for all impact assessments for civil engineering and planning.

Foraminifers (*Baculogypsina sphaerulata*), also called "star sand" because some beaches (such as one in Okinawa, Japan) have sand composed almost exclusively of this millimetric microfossil.

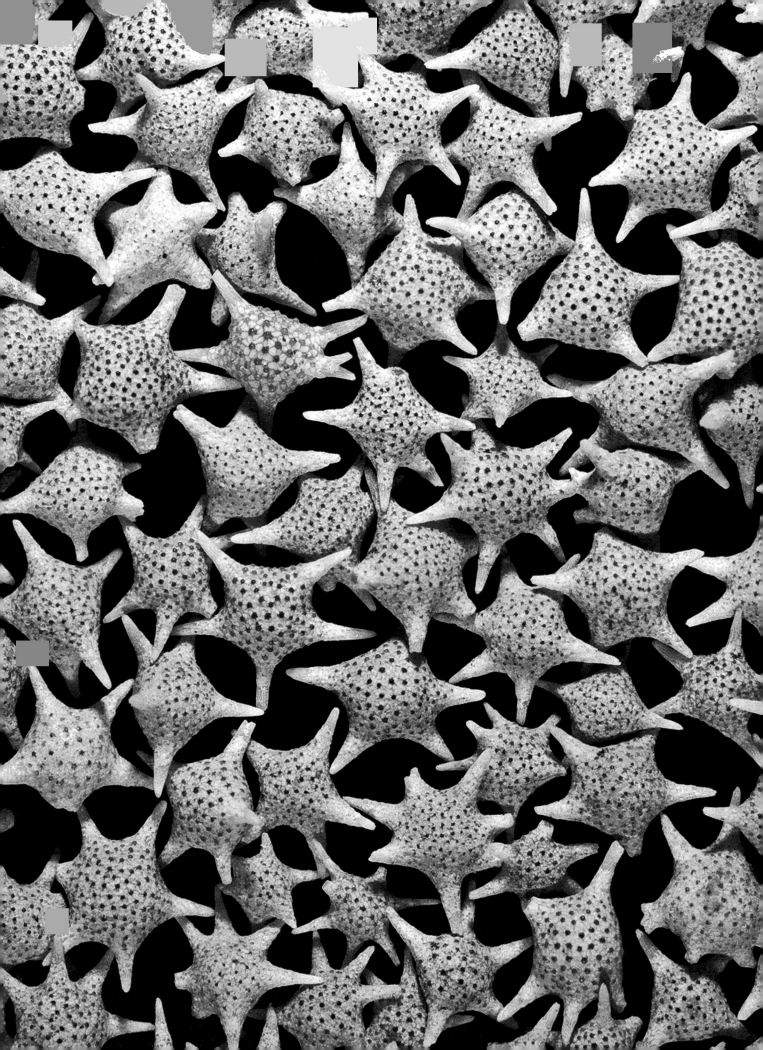

Microfossils
for Conducting Investigations

In addition to consulting frequently with the customs authorities that govern the import and export of sensitive items, paleontologists are also sometimes called upon in legal and police cases. From small fragments of limestone, marl or mud stuck in shoe soles or the organisms contained in the water of a drowned person's lungs, it is possible to provide information regarding the origin of the sediment or the water. Pollen grains, spores, diatoms and so on can all be pieces of evidence. For example, in the Bruay-en-Artois murder case in France in the 1970s, fossils found in the notches of shoe soles were what proved that certain individuals were present in a specific place.

In the case of theft, forensic paleontological examinations are delicate and fascinating. They have been conducted when precious objects have been stolen and replaced with rocks during transport. For example, containers of perfumes once arrived at their destination filled with sand consisting of organisms known to exist only on some beaches in warm regions, and small boxes of diamonds were replaced by limestone sometime between their departure from South Africa and their arrival in Antwerp, Belgium. In each instance, the same approach was applied by using two different sources of information: the identity of the rock itself and its surface state. The block of rock, the only piece of evidence, revealed its geologic origin and thus the location where the swap took place. For thieves, a stone is only a stone; it cannot speak. As the saying goes, it's "dumb as a rock." Next, the information that could be learned from the rock's surface was examined. Indeed, the existing bioindicators, such as pollen grains, took the reins. Studying the rock's surface made it possible to find environmental indications of where the rock was collected prior to the swap.

Gymnosperm pollen grain (tetrad, *Classopollis* sp.) from rocks that are approximately 120 million years old (Early Cretaceous). Photo taken with a scanning electron microscope and colorized.

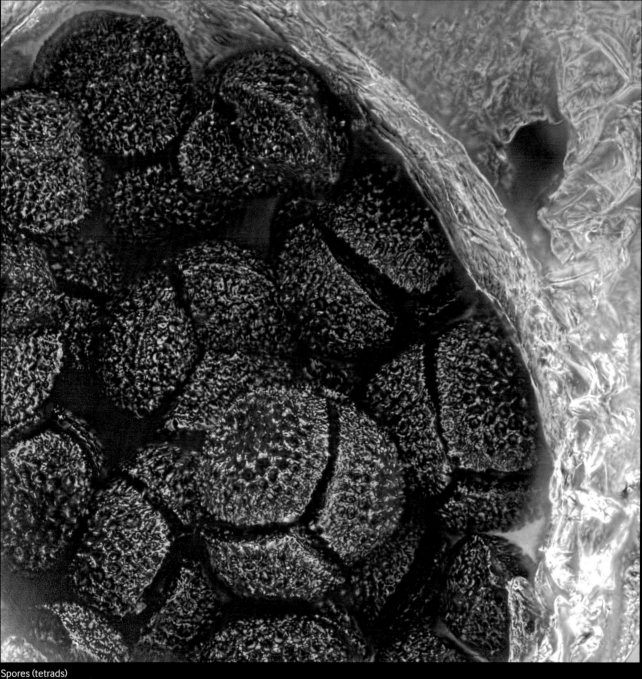

Spores (tetrads) in sporangium of present-day *Riccia sorocarpa*. Photo taken with an optical microscope.

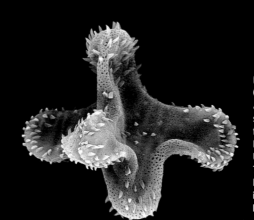

Pollen grain (*Aquilapollenites attenuatus*). The specimen comes from rocks in North Dakota (Hell Creek, USA) that are approx. 70 million years old. It measures approximately 0.05 mm in length.

Without the sun reflecting
What need exists for crystal?
Such a small bit of glass
Of ingenious carpentry
For crossing the expanses
Of sidereal time.

Jean-Yves Reynaud

This Cross, Our Sacred Banner
for Rock Dating

The shells of fossil radiolarians are made of rock crystal. Some have four arms that are made of a multilayer network, and each individual type has its own geometry. Sometimes longitudinal bars are connected by thin rods; other times a three-dimensional interlacing of bars gives the shell a massive spongy look. This structure protects the inner cytoplasm. A lightweight network allows the cytoplasm to increase its apparent volume by forming foam with seawater (a little cytoplasm, a lot of water, like how beer foam is made: a little liquid, a lot of air!). The spines serve as a support for the long outward extensions that allow the flagella to capture food.

These superbly perforated crosses measure approximately half a millimeter. They are not on the royal pendant of some Lilliputian queen but rather are the skeletons of fossil radiolarians several millions of years old.

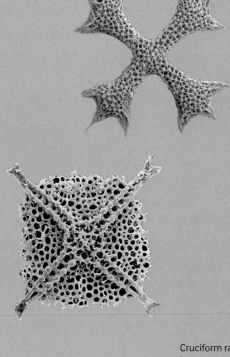

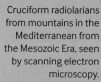

Cruciform radiolarians from mountains in the Mediterranean from the Mesozoic Era, seen by scanning electron microscopy.

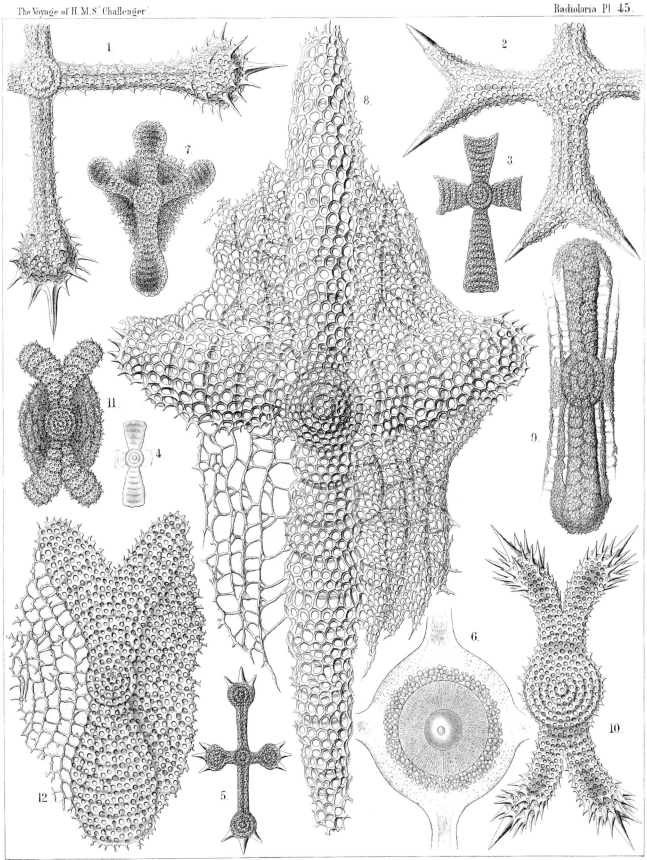

E Haeckel and A.Giltsch.Del.

L.Giltsch Jena. Lithogr.

1-6.HAGIASTRUM, 7-9.HISTIASTRUM, 10.AMPHIRHOPALUM.
11.12.AMPHICRASPEDUM.

Very Modest Sizes
for Great Achievements

Looking at the microscopic world reveals very well-organized structures, both in terms of their overall structure and their smaller detailed parts. These structures result from millions of years of "tests" of their resistance as much as of maximizing their resources. The resulting forms are a response to physical and chemical needs, yet everyone agrees that they are particularly beautiful. It is hardly surprising that these structures have inspired architects or that they invite us to build a bridge between art and science.

The skeletal structures of tiny organisms have inspired the great ambitions of architects. La Géode in the park at the Cité des Sciences near the Porte de la Villette in Paris was made by assembling a multitude of bars into triangles to form hexagons, just like radiolarians create their mesh-like shells.

La Géode (Parc de la Villette, Paris). La Géode's design (36 m in diameter) started with an icosahedron whose edges were each divided into 10. It is composed of 6,433 equilateral triangles. In 1983 French architect Adrien Fainsilber imagined and designed La Géode. It would later be built by Gérard Chamaillou in 1986.

Spumellarian (radiolarian) skeleton from rocks in Deva, Romania, dating from the Late Cretaceous Period (approx. 80 million years ago). Size: approx. 0.3 mm.

A Pretty Rosette
for the Architect

Nature inspires dreams, but it also has qualities on a variety of scales that people seek to adapt. This transfer is called biomimetics, or bionics.

The frustule wall of a diatom (*Pleurosigma*, for example) looks flattened and slightly elongated, like the filling between the gaps of a layer of contiguous vesicles. It has a high level of stability and served as a model for two German engineers: Frei Otto, for the roof of the Olympic Park in Munich for the 1972 Olympic Games, and Markus F. Manleitner, for constructing his Hippopotamus Houses at the Berlin Zoo.

Diatoms are sometimes abundant in oceans. When they accumulate on the seabed, they produce siliceous sediments that are sometimes rich in oil and gas (such as the diatomite in Monterey, California, or in Pisco, Peru, and which at the surface burns naturally, as can be seen in the southern suburbs of Los Angeles).

The porosity of diatomite makes it a light rock, a very useful characteristic for architecture. The dome of the Hagia Sophia in Istanbul is made of diatomite, the only rock light enough for such a size.

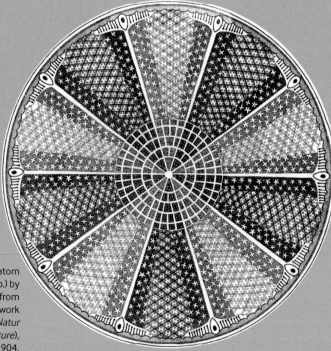

Illustration of a diatom
(*Thalassiosira* sp.) by
E. Haeckel from
his published work
Kunstformen der Natur
(*Art Forms in Nature*),
Plate 4, 1899–1904.

Dome of the Hagia Sophia of Constantinople (Istanbul, Turkey), seen from the inside. Its geometry evokes that of diatoms.

Roof of the Munich stadium for the Munich Olympic Games in 1972, by the architect Frei Otto.

Index

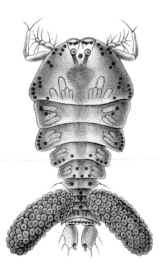

Geologic Time Scale

Eon	Era	Period	Epoch	Age (Ma)
Phanerozoic	Cenozoic	Quaternary	Holocene	present 0.0117
			Pleistocene	2.588
		Neogene	Pliocene	5.333
			Miocene	23.03
		Paleogene	Oligocene	33.9
			Eocene	56.0
			Paleocene	66.0
	Mesozoic	Cretaceous	Late	100.5
			Early	~145.0
		Jurassic	Late	163.5±1.0
			Middle	174.0±1.0
			Early	201.3±0.2
		Triassic	Late	~235
			Middle	247.2
			Early	252.2±0.5
	Paleozoic	Permian	Lopingian	259.9±0.4
			Guadalupian	272.3±0.5
			Cisuralian	298.9±0.2
		Carboniferous	Pennsylvanian Late	307.0±0.1
			Pennsylvanian Middle	315.2±0.2
			Pennsylvanian Early	323.2±0.4
			Mississippian Late	330.9±0.2
			Mississippian Middle	346.7±0.4
			Mississippian Early	358.9±0.4

Eon	Era	Period	Epoch	Age (Ma)
Phanerozoic	Paleozoic	Devonian	Late	358.9±0.4
			Middle	382.7±1.6
			Early	393.3±1.2
		Silurian	Pridoli	419.2±3.2
			Ludlow	423.0±2.3
			Wenlock	427.4±0.5
			Llandovery	433.4±0.8
		Ordovician	Late	443.4±1.5
			Middle	458.4±0.9
			Early	470.0±1.4
		Cambrian	Furongian	485.4±1.9
			Epoch 3	~497
			Epoch 2	~509
			Terreneuvian	~521
				541.0±1.0

Eon	Eon	Era	Age (Ma)
Precambrian	Proterozoic	Neoproterozoic	541.0±1.0
		Mesoproterozoic	1000
		Paleoproterozoic	1600
	Archean	Neoarchean	2500
		Mesoarchean	2800
		Paleoarchean	3200
		Eoarchean	3600
	Hadean		4000
			~4600

Ma = mega-annum (millions of years)

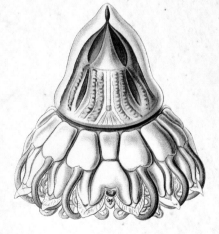

Photo Credits

b: bottom
r: right
l: left
t: top
m: middle
a: all

Charles Bachy: 125
Richard Bartz: 167
M. Beauregard: 65 (b)
Karim Benzerara: 63 (b)
Antoine D. Bercovici, PhD: 245 (b)
Daniel Bernoulli: 89 (br), 178
L. P. Bert: 29 (b)
Marie-Madeleine Blanc-Valleron & Annie
    Cornée: 17
Pierre Bultynck: 143
Patrick Cabrol: 59
M. Carmichael: 111 (b)
Sylvain Charbonnier: 25 (m)
Clem: 239
Jean-Paul Colin: 133-135
Alain Couté: 87 (t, br), 191 (b), 201
CSIRO: 23 (t), 91 (t), 93 (t), 112 (l), 113 (b)
M. Dagnino: 208
Paolo da Reggio: 215 (tr)
Jean Dejax: 113 (t), 165, 244
Frédéric Delarue: 63 (t, m)
Des Callaghan: 245 (t)
Patrick De Wever: 10 (tl), 33, 45, 64, 68, 155,
    163, 184, 196, 203 (b), 205, 206
Didier Descouens: 62
DIMSFIKAS: 249 (t)
Robert Doisneau: 23 (b)
John Dolan: 21 (t), 47, 81, 131 (b), 148-149
DR: 20 (b), 21 (bl, br), 183 (t)
Paulian Dumitrică: 9 (2nd row l, bl), 19, 35,
    41-43, 86, 87 (ml, bl), 88, 89 (bl), 101, 107,
    109 (tm), 124, 138 (t), 145, 155, 162, 172,
    174, 202, 203 (t), 241, 249 (ml)
Electron and Confocal Microscopy
    Laboratory: 9 (tr)
EMPTiness: 181 (tl), 183 (b)
Bernard Faye/MNHN: 207
Mike Foster: 179-180, 181 (all except tl)
Bernard Gagnon: 214, 215 (t)
Parent Géry: 69 (b)
R. Gieseke: 118 (r)
Giomodica: 249 (b)
Hélène Grenier: 209 (r)
Hannes Grobe: 9 (tl, 3rd row l), 10 (tm, tr, bl),
    11, 105, 109 (tr), 185, 236-237
G. Hallegraeff: 167 (m)
Roy Halling: 108
P. Harrison: 65 (t)
Don Hitchcock: 51
Dr René Hoffmann: 69 (t)

Pr. S. Imoto: 89 (t)
Emmanuelle Javaux: 61
Jmpost: 93 (b), 106
Kvikende: 70
Philippe Loubry: 16, 20 (t)
Magdalena Lukowiak: 109 (tl), 138 (all
    except t), 170, 191 (t)
Edwige Masure: 112 (b)
Atsushi Matsuoka: 25 (t)
Michael: 9 (br)
Christopher Michel: 71
Anatoly Mikhaltsov: 199
Jean-David Moreau: 25 (b)
David Moreira: 193 (t)
Musée du Sable: 243
Mwanasimba: 209 (l)
Myrabella: 251 (t)
NASA Jesse Allen & Norman Kuring: 97
NASA Goddard: 96
NASA Steve Groom Askewmind: 27
Marie-Lan Nguyen: 90
Simona Pestrea/MNHN: 2-3 (Deflandre
    collection), 254 (Tempere collection)
Picturepest: 200 (t)
Pixelto: 9 (2nd row r)
Poniol: 215 (br)
Francis Robaszynski: 238
Jorge Royan: 251 (b)
Jean-Paul Saint-Martin: 193 (bl, br)
Arito Sakaguchi & IODP/TAMU: 40
Christian Sardet: 67, 111 (t), 197
Christian & Noé Sardet: 95
Thomas Servais (CNRS, University of Lille):
    120
ShavPS: 240
Skouame: 29 (t)
James St. John: 73
Niels Swanberg: 75, 79, 109 (b)
U.S. Department of Agriculture: 22
Thijs Vandenbroucke (CNRS, University of
    Lille): 121 (t)
Jean Vannier: 144
Marco Vecoli (CNRS, University of Lille):
    121 (b)
Verisimilus: 147 (t)
Matt Wilson/Jay Clark: 118 (l)
Jakub Witkowski: 115, 131 (t)
Shuhai Xiao: 76-77
Z. Xiguang: 147 (b)
Jeremy R. Young: 167 (b), 168-169

254

MARVELOUS MICROFOSSILS

Acknowledgments

During the writing of this book I benefited from the graciousness of many colleagues, for photos, advice or help. I am grateful to them for that. As small streams make big rivers, the creation of this book was possible thanks to their contributions. Their names in alphabetical order are: Charles Bachy, Karim Benzerara, Daniel Bernoulli, Marie-Madeleine Blanc-Valleron, Patrick Cabrol, Margaux Carmichael, Laurent Carpentier, Sylvain Charbonnier, Annie Cornée, Jean-Paul Colin (†), Alain Couté, Sylvie Crasquin, Jean-Claude Daniel, Frédéric Delarue, Jean Dejax, Claire Dericke, Francine Derouville (Atelier Doisneau), John Dolan, Grégoire Egoroff, Michael Foster, Béatrice Garnier, Spela Gorican, Gustaaf Hallegraeff, Emmanuelle Javaux, Jean-Yves Kernel, Nathalie Legrand, Alexandre Lethiers, Philippe Loubry, Magdalena Lukowiak, Atsushi Matsuoka, Denis Moreau, Didier Néraudeau, Luis O'Dogherty, Lisa Price, Francis Robaszynski, Christian Sardet, Thomas Servais, Thijs Vandenbroucke, Jean Vannier, Marco Vecoli and Anne Versini.

A special thank-you to my old friend Paulian Dumitrică for his many, many pretty pictures and his invaluable assistance.

Bibliography

Bignot, Gérard. *Introduction à la Micropaléontologie* (*Elements of Micropalaeontology*). Dunod, 2001.

De Wever, Patrick. *Le beau livre de la Terre* (The illustrated book of the Earth). Dunod, 2014.

De Wever, Patrick, Bruno David and Didier Néraudeau. *Paléobiosphère, Regards croisés des sciences de la vie et de la Terre* (Paleobiosphere: Life and earth science perspectives). Vuibert, 2010.

De Wever, Patrick, and Karim Benzerara. *Quand la vie fabrique les roches* (When life makes rocks). EDP Sciences, 2016.

Haeckel, Ernst. *Die Radiolarien* (*Rhizopoda radiaria*): *Eine Monographie* (Radiolaria [*Rhizopoda radiaria*]: A monograph). Berlin: Verlag von Georg Reimer, 1862.

Haeckel, Ernst. *Report on the Scientific Results of the Voyage of H.M.S.* Challenger *during the Years 1873–76.* Edinburgh: Adam & Charles Black, 1887.

Haeckel, Ernst. *Kunstformen der Natur* (*Art Forms in Nature*). Leipzig and Vienna: Verlag des Bibliographischen, 1899–1904.

Haq, Bilal U., and Anne Boersma, eds. *Introduction to Marine Micropaleontology.* Elsevier, 1978.

Lipps, Jere H., ed. *Fossil Prokaryotes and Protists.* Blackwell Science Inc., 1992.

Sardet, Christian. *Plancton: Aux origines du vivant* (*Plankton: Wonders of the Drifting World*). Ulmer, 2013.

This edition published by arrangement with Biotope éditions, Mèze, France.

First published in France under the title *Merveilleux microfossiles. Bâtisseurs, chronomètres, architectes* by Biotope éditions, Mèze. © Biotope éditions, Mèze, France 2016.

Johns Hopkins University Press
2715 North Charles Street
Baltimore, Maryland 21218-4363
www.press.jhu.edu

Library of Congress Cataloging-in-Publication Data

Names: De Wever, Patrick, author.
Title: Marvelous microfossils : creators, timekeepers, architects / Patrick De Wever ;
    foreword by Hubert Reeves ; translated by Alison Duncan.
Other titles: Merveilleux microfossiles. French.
Description: Baltimore : Johns Hopkins University Press, [2020] |
    Originally published in France under the title Merveilleux
    microfossiles. Bâtisseurs, chronomètres, architectes by Biotope
    éditions in 2016. | Includes bibliographical references and index.
Identifiers: LCCN 2019023269 | ISBN 9781421436739 (hardcover : alk. paper) |
    ISBN 1421436736 (hardcover : alk. paper) | ISBN 9781421436746 (electronic)
    ISBN 1421436744 (electronic)
Subjects: LCSH: Micropaleontology.
Classification: LCC QE719 .D4913 2020 | DDC 560—dc23
LC record available at https://lccn.loc.gov/2019023269

A catalog record for this book is available from the British Library.

Special discounts are available for bulk purchases of this book. For more information, please contact Special Sales at 410-516-6936 or specialsales@press.jhu.edu.

Johns Hopkins University Press uses environmentally friendly book materials, including recycled text paper that is composed of at least 30 percent post-consumer waste, whenever possible.